Noise in the Plastics Processing Industry

Second Edition

Bob Peters

CRC Press
Taylor & Francis Group
Boca Raton London New York

CRC Press is an imprint of the
Taylor & Francis Group, an **informa** business

First published 2017 by Smithers Rapra

Second edition published by CRC Press
Taylor & Francis Group
6000 Broken Sound Parkway NW, Suite 300
Boca Raton, FL 33487-2742

First issued in paperback 2020

ISBN 13: 978-0-367-65700-0 (pbk)
ISBN 13: 978-0-367-03025-4 (hbk)

Visit the Taylor & Francis Web site at
http://www.taylorandfrancis.com

and the CRC Press Web site at
http://www.crcpress.com

Preface

Noise in the Plastics Processing Industry: A Practical Guide to Reducing Noise from Existing Plant and Machinery was published by RAPRA in 1985. It was written by a group of experts in noise and vibration, including several from the Health and Safety Executive, with detailed knowledge of the industry. It was 52-pages long, with 25 pages devoted to 25 case studies, illustrating successful examples of noise reduction. The purpose of the book was to provide practical guidance to reducing noise at the source to reduce the noise exposure of those employed in the industry, and preventing noise-induced hearing loss in employees.

The risk of damage due to exposure to high levels of noise has been known about for a long time. In 1713, a treatise on diseases in workers by the Venetian physician B. Ramazzini included a description of how those hammering copper have their ears so injured by that perpetual din …. '*that workers of this class become hard of hearing and, if they grow old at this work, completely deaf*'. With the coming of the industrial revolution and the introduction of steam power, the problem of occupational noise exposure became acute. Workers making steam boilers in particular succumbed to deafness in such numbers that it became known as 'boilermakers disease'.

In 1831, Dr J. Fosbroke, writing in the *The Lancet* claimed that deafness in blacksmiths was a result of their occupation. In 1886, a Glasgow physician, Dr. T. Barr, found that ≈75% of the boilermakers that he examined were deaf to the extent that they had difficulty in hearing (or could not hear at all) a public speaker. He compared these boilermakers with some in other occupations and found, for example, that only 8% of letter-carriers suffered deafness to the same degree.

In 1963, the *Wilson Report*, a widespread report on all aspects of noise affecting society, recommended three lines of attack on the problem of hearing loss from noise exposure in the workplace: more widespread voluntary action within industry on the basis of existing knowledge; legislation on the basis of existing knowledge; research to try to obtain a more definite understanding of the relationship between noise and hearing loss, with legislation to follow, if necessary, when the results of the research are available.

Also in 1963, the UK government publication *Noise and the Worker* was issued to provide specific guidance concerning the hazards of occupational noise, and the date of issue of this booklet became, by general consensus, the time from which employers should have been aware of the risks, and taken actions to protect their employees.

In 1972, the UK government issued a code of practice for reducing the exposure of employed persons to noise which contained guidelines for noise-exposure limits framed in terms of a newly defined noise measure, the continuous equivalent noise exposure level over a duration of 8 h: LAeq,8 h. In 1981, draft regulations were published in response to an European Community (EC) directive and in response the UK government issued draft proposals for legislation in a consultative document *Protection of Hearing at Work*, which eventually led to the *Noise at Work Regulations 1989*.

In 1970, the Health and Safety Executive (HSE) published a guidance document entitled *Reduction of Machinery Noise*.

In 1982, the HSE issued the first of several collections of Case Studies *100 Practical Applications of Noise Reduction Methods*.

The 1985 RAPRA guide was published in response to this guidance – this volume is an updated version of the 1985 guide *Noise in the Plastics Processing Industry: A Practical Guide to Reducing Noise from Existing Plant and Machinery*.

So what has changed since 1985? It is perhaps easiest to start first with what has not changed. The laws of physics that determine how sound and noise is generated and propagates from the source to the receiver have not changed. Thus, breaking up lumps of hard plastic with rotating steel-cutting blades remains (as in 1985) an inherently noisy process. Human hearing has not changed: prolonged exposure to high levels of noise in the workplace continues to cause noise-induced hearing loss, or industrial deafness. The standard methods for reducing noise still involve, as in 1985, using sound-insulating, sound-absorbing and vibration-isolating materials, and the use of standard noise-control solutions such the use of enclosures, silencers screens and barriers.

One new noise-reduction technique has been developed – active noise control – but it is more suited to reduction of a fairly constant level of low-frequency noise in a specific location rather than variable levels of high-frequency machinery noise, although it can be used for the reduction of noise from fans and compressors.

Other changes include:

• Improvements in noise prediction, including 'noise mapping' software.

- Improvements in instrumentation to give better diagnostic techniques.

- Advances in technology has meant that many noise-producing processes and activities can be automated so that employees can monitor and control machinery remotely from a quiet environment, rather having to be in attendance close to the machine and being exposed to high noise levels.

- New quieter technologies have replaced older noisier processes e.g., replacing petrol- or diesel-powered equipment by much quieter electric motors; replacing impact forces by hydraulically/pneumatically driven silent 'squeeze' forces.

- Vastly increased amount of information published and readily available about noise and its control.

- More regulations and guidance documents have been published and come into operation since 1983. These include the European Union (EU) Directives on *Noise in the Workplace*, and on *Machinery Noise*; UK *Noise at Work Regulations* in 2005; and the *HSE Noise Case Studies* on noise reduction measures.

Nobody reads a book like this from cover-to-cover like a 'page turner' novel – rather they will read certain chapters, or sections of chapters depending on their interest and their 'need to know' at the time. Therefore, an attempt has been made to help readers to navigate through the book and find what they want to know. In addition to a detailed contents list at the front of the book, each chapter starts with an introduction explaining what the chapter will cover, and concludes with a brief summary.

Chapter 1 covers a brief explanation of the ideas and terminology which will be used in later chapters. **Chapter 2** is a review of the ways in which noise can be reduced either at the source, during the sound transmission path, or at the point of reception. **Chapter 3** gives examples of how noise from plastics processing machinery can be reduced with material taken from the 1985 RAPRA publication. **Chapter 4** covers the nature of noise-induced hearing loss in the context of noise exposure levels in the workplace. The requirements of the *Noise at Work Regulations 2005* are discussed in terms of noise-exposure action levels and limits which, if exceeded, impose duties on employers and employees. **Chapter 5** discusses the types of hearing protection available, their performance, issues relating to their selection and use as part of a hearing-conservation programme (including difficulties and limitations) and the need for information instruction and training for employees. **Chapter 6** covers noise generated in the workplace which can also cause disturbance to people in the vicinity. This chapter briefly reviews target noise levels which should be achieved within the environment so as not to cause disturbance to people living and working nearby. It outlines strategies for minimising noise emission and suggests relevant standards and codes of practice relating to industrial noise in the environment and noise disturbance. **Chapter 7** discusses the ability to predict noise levels and how they can play an important part in the control of noise. The methods used to predict noise levels within and outside the workplace are briefly outlined and the benefits of prediction of noise levels illustrated using a simple model. **Chapter 8** shows the different ways of specifying noise emission from machinery, and the EU regulations on machinery noise. Noise test codes for plastics granulators and shredders are reviewed and a report on the effectiveness of directives is discussed. **Chapter 9** discusses that, although the priority should always be given to using quieter machinery and equipment wherever possible, much can also be done to reduce noise levels by careful design of the workplace. **Chapter 10** lists various case studies. A very important part of the original 1985 book was the use of case studies to illustrate the various ways in which noise has been reduced in workplaces in the plastics processing industry. Since then, case studies for all types of noisy industries have been widely publicised by the HSE. This chapter reviews some existing case studies, including those in the original 1985 RAPRA book, and presents some new ones.

The original book, published in 1985, was practical in its approach, with limited theory and no mathematical equations. As far as possible, the same approach has been adopted in this version, with the justification that there are now many other texts available where the reader can find more detailed theory and mathematical prediction formulae if required. A bibliography of further reading is given at the end of each chapter.

The 1985 RAPRA book will be a hard act to follow, but it is hoped that readers will find this revised version informative and useful.

Contents

1 Basic Concepts and Terminology of Sound and Vibration

1.1 Sound Pressure, Frequency and Wavelength

1.1.1 Sound

Humans can hear sounds that are very small and rapid fluctuations which occur in air pressure that can be detected by our eardrums. However, how small are the sounds and how rapid are the fluctuations?

To answer the first question we must compare the size of the fluctuations with the typical atmospheric pressure. To answer the second question, 'how rapid' is related to the idea of the frequency or pitch of a sound.

1.1.2 Sound or Noise?

A very common definition of noise is 'sound that is unwanted by the recipient'. At home or at leisure, we can distinguish between sounds that we like and want to hear or listen to, and those we do not, which we call 'noise'. The difference is very subjective, so what is a wanted and enjoyable sound for some people is regarded as noise that is unwanted for other people.

If, however, we are considering occupational noise (i.e., noise in the workplace) and the possibility of noise-induced hearing loss, the distinction disappears because the damage to the hearing system may occur whether it is caused by sounds from noisy machines or from prolonged exposure to very loud music.

1.1.3 Sound Pressures

A typical value of atmospheric pressure is about 100,000 Pascals (Pa), but varies with weather conditions. 1 Pa = 1 Newton per square metre (N/m^2), the scientific unit of pressure. In more everyday terms, this is about 15 lbs force per square inch, or 1 kg force per square centimetre.

The size of the fluctuations in atmospheric pressure which (if rapid enough) create the sensation of sound range are from about 2 millionths of 1 Pa to about 200 Pa. The lower end of this range is the threshold of hearing and the upper limit is the loudest sound that would be tolerable to the human ear – the threshold of pain. Both thresholds vary from person to person.

Hence, even the loudest sound in the audible range is only about one-thousandth of the everyday, more or less steady atmospheric pressure on human eardrums. The range of pressures, from smallest to largest, is about 5 million to 1: a very large range. The amount of energy in a sound is proportional to the square of the sound pressure. Hence, when the range of sound pressures is translated to a range of sound energies, it becomes 25 million million to one: a truly enormous range. This makes us realise what how delicate and sensitive our ears are as detectors of sound.

This enormous range of sound pressure and sound energies in the audible range of sounds is one of the reasons that sound pressure is usually measured using a logarithmic scale called the decibel (dB) sale, which helps to 'squash' or compress the range in a more manageable and convenient range. The dB scale is explained later in this chapter.

1.1.4 Frequency

The rapidity of pressure fluctuations at the human ear which cause the sensation of sound is expressed as the frequency of the sound, and is measured in cycles per second, or Hertz (Hz). The range of sound frequencies which are audible to the human ear extends from about 20 Hz to about 20,000 Hz [i.e., 20 kilo Hertz (kHz)] but varies from person to person and also depends upon age. The subjective sensation of the 'pitch' of a musical note is closely related to its frequency.

1.1.5 Broadband Noise

Most sounds and most noises contain a wide variety of frequencies, and are often referred to as 'broadband' noises. Broadband sounds which contain a more or less even spread of frequencies across the audio range include: 'hissing'-type sounds; the noise of a boiling kettle, from a frying pan, from a fountain or waterfall, from an FM radio off-tune in between stations; 'white' and 'pink' noise used in audio testing.

Noise from an electric drill and many other cutting and abrasive power tools generates mainly high-frequency noises, mainly in the 2000, 4,000 and 8,000 octave bands, and noise in industry generally tends to be in the range of 3,000–6,000 Hz. At these high frequencies, the human ear is most sensitive and most susceptible to damage from over exposure to high levels of noise. Noise from pumps, fans, and diesel engines tends to have a frequency spectrum (range of frequencies) with the highest levels at lower frequencies, often in the range 63 to 250 Hz.

1.1.6 Pure Tones

A sound of just one single frequency is called a 'pure tone'. The time signal 'pips' on the radio and computer beeps are examples of a pure tone. A note played on a piano or guitar is a mixture of pure tones and harmonics (multiples of the tone frequency). Rotating machinery such as fans, motors, and gears produce pure tones and are characterised by descriptors such as a 'whine', 'whistle', 'drone' and 'hum'. Sounds with a tonal component are important in the assessment of noise in the environment because they are considered to be more annoying than similar levels of broadband noise without the pure tone. Some machines and fans, for example, generate noise with a mix of broadbands and tones.

1.2 Frequency Weightings and Frequency Analyses

The frequency content of a sound can be measured and investigated in one of two ways. The first way involves applying an electronic frequency weighting (A, C or Z), which gives a simple and rapid method based on a single measurement. The second way is the more complicated method of frequency analyses, which splits the sound into various bands of frequencies, usually into octaves or 1/3 octave bands.

1.2.1 A-Weighted Decibels

The vast majority of noise measurements are of A-weighted decibels (dBA). A-weighting is an electronic frequency-weighting network that attempts to build the human response to different frequencies into the reading indicated by a sound-level meters so that it relates to the loudness of the sound.

1.2.1.1 A Brief History of the A-Weighting and its Relationship to Loudness

Although the human ear can detect sound waves over a wide range of frequencies, from about 20 Hz to about 20 kHz, it is not equally sensitive to all these frequencies. It is most sensitive to frequencies in the range 1,000–4,000 Hz (i.e., 1–4 kHz). This is also the range at which most noise often occurs in industry and the range at which the ear is most sensitive to damage from over exposure to noise (i.e., to noise-induced hearing loss, or 'industrial deafness'). The human hearing system is less sensitive to lower frequencies (below about 5,000 Hz).

The variation in human hearing to different frequencies was investigated by Fletcher and Munson (1933) in the USA and by Robinson and Dadson (1954) in the UK, among others. They tested how human subjects judged the loudness of pure tones at different sound levels and frequencies. This work resulted in the production of a set of 'equal loudness contours'. These showed, for example, that a tone at a low frequency, such as 100 Hz, would have to be several dB higher to sound equally as loud as a tone at a higher frequency of 1,000 Hz.

The shape of these curves was simulated electronically and built into the sound-level meter to give a sound-level reading, in dBA, which to some extent at least corresponded to the sensation of the loudness of the sound.

Originally, three frequency weightings were produced: A, B and C. These correspond to the difference in the shapes of the equal loudness contours at different sound levels. The B weighting is now no longer in use and the A weighting has been used throughout the world as the usual way to measure sound and noise levels. The C weighting was designed originally to simulate the human hearing response at high levels of sound. The C weighting is much flatter (i.e., shows much less variation with frequency) than the A-weighting curve and is still used (e.g., in the *Noise at Work Regulations*) if a more or less flat, unweighted frequency response is required. In 2003, the Z-weighted decibel (dBZ) was introduced. This is completely flat and will probably eventually replace the use of the C weighting. The values of these weighting (to the nearest dB) at octave-band frequencies are shown in **Table 1.1**.

Table 1.1 Frequency weightings (to the nearest dB)			
Octave band (Hz)	dBA	dBC	dBZ
63	−26	−1	0
125	−16	0	0
250	−9	0	0
500	−3	0	0
1,000	0	0	0
2,000	1	0	0
4,000	1	−1	0
8,000	−1	−3	0
dBC: C-weighted decibel			

Table 1.2 Some typical sound levels in dBA	
140 dBA	Threshold of pain
120 dBA	Jet aircraft at 100 m
100 dBA	Very loud road drill, loud disco
90 dBA	DIY drill (close to ear), food blender, lorry (roadside)
80 dBA	Alarm clock, doorbell, telephone ringing, traffic at a busy roadside
70 dBA	Moderately loud hair dryer, vacuum cleaner
60 dBA	Normal person–person conversation, washing machine
50 dBA	Quiet urban daytime, large office, TV in domestic living room
40 dBA	Quiet office, library
30 dBA	Bedroom at night, soft whisper
20 dBA	Just audible, broadcasting studio, rustling leaves
0 dBA	Threshold of hearing (for a young person with normal hearing)

1.2.2 Frequency Analyses

Although a single value such as dBA is useful for assessing the effect of a noise, a more detailed knowledge of the frequency content of the noise is sometimes required, for example, if it is necessary to specify noise-control measures. The sound pressure level(s) (L_p) is measured in a series of frequency bands and shown as a graph or chart of sound pressure levels against frequency, called a frequency spectrum. The most commonly used method of frequency analysis is octave-band analysis, which is appropriate for noise-control purposes. 1/3 octave-band analysis is used for measurements of building acoustics and narrowband analysis is used to identify the frequency and thence the source of pure tones among broadband noise.

Examples of octave-band noise spectra are shown in **Figure 1.1** for a typical predominantly low-frequency noise such as a fan or a pump, for a predominantly high-frequency noise such as an electric drill or angle grinder, and also for pink noise, which has a flat noise spectrum. Although the overall dBZ levels are similar for low- and high-frequency sources (81 and 82 dBA, respectively) there is a big difference in their dBA levels: 82 dBA for the mainly high-frequency noise source and 69 dBA for the low-frequency source. For a source of pink noise with a constant octave-band level of 65 dB, the overall noise levels are 72 dBA and 74 dBZ, respectively.

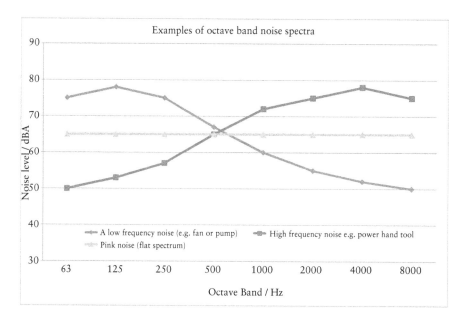

Figure 1.1 Examples of octave-band noise spectra

Sometimes, if the noise contains a tone, or series of tones, it is better to carry out a narrowband analysis to identify the exact frequencies of the various tones. This strategy can help to identify, rather like a fingerprint, the particular source of the tonal noise. **Figure 1.2** shows an example of a narrowband analysis.

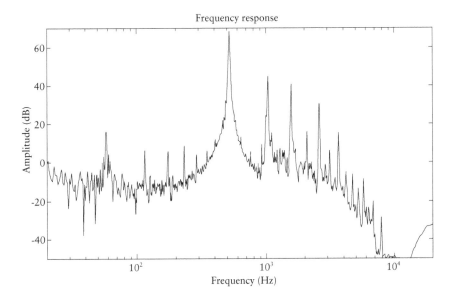

Figure 1.2 Example of a narrowband analysis of a tonal sound. Note that there are several examples of narrowband spectra in the case studies of **Chapter 10**. Reproduced with permission from E. Ballastero, London South Bank University, London, UK. ©2017, London South Bank University

1.2.3 Wavelength

At any point in a sound wave, the sound pressure varies periodically with time at a rate which is dependent on the frequency. At a nearby point, the sound pressure will also vary in the same way, but at the same moment will be slightly out-of-step, or out-of-phase, with the pressure at the first point. As the distance between the two points increases, their sound-pressure cycles become increasingly out of step. That is, their phase difference increases until the distance between the points reaches a certain value, and then the pressure fluctuations at the two are back in step, back in phase. This distance is the wavelength of the sound at that frequency. Looking at waves on the surface of water, the wavelength is the distance between two wave crests, or between two wave troughs.

The frequency, f, and wavelength, λ, of the tone are related by the speed of sound, c, in the medium according to **Equation 1.1**:

$$c = f.\lambda. \tag{1.1}$$

The speed of sound in air varies with temperature, from about 330 m/s at 0 °C to about 340 m/s at 20 °C. The speed of sound in air does not vary with frequency, and so the wavelength decreases as the frequency increases. Hence, low-frequency sound has a long wavelength and high-frequency sound a short wavelength. Taking the latter figure for the speed of sound in air, **Equation 1.1** gives a wavelength of 0.34 m at 1,000 Hz, increasing to 3.4 m at 100 Hz and reducing to 0.034 m (or 3.4 cm) at 10 kHz.

1.2.3.1 The Significance of Frequency and Wavelength in Noise Control

The velocity of sound depends only on the medium in which the sound is propagating, for example, air or water, or some building material, such as brick or concrete. The frequency of the sound depends entirely on the sound source. A sound wave may be transmitted from air *via* brickwork into water and then back into air again – and its frequency would remain unchanged as it moved from one medium to another, even though the wavelength would change.

It is important to know the frequency content of sounds because almost all of the acoustic parameters and properties associated with noise measurement and control vary with frequency, including the sensitivity of human hearing, the sound-absorbing and sound-insulating properties of materials, and the performance of noise sources.

Thus, for example, the effectiveness of various noise-control measures such as attenuators and silencers, acoustic tiles, partitions, enclosures, barriers and anti-vibration mounts varies with frequency. Usually, it is easier to achieve noise reductions for high frequencies than for lower frequencies. The subjective impression created by the sound will also depend very much on its frequency content. Pure tones tend to be more annoying than broadband noise at the same sound level. The frequency of a pure tone also determines the subjective nature of its pitch. The frequency of a tone also determines the subjective measure of its pitch.

Wavelength is important in determining how sound travels. That is, how it is reflected by surfaces, scattered by obstacles, and the extent to which it bends around barriers (a property of sound called 'diffraction'). The wavelength of a sound has a profound effect on the way it interacts with obstacles in its path. Thus, for example, the effectiveness of a noise barrier in shielding noise depends on its size compared with the sound wavelength, and so the barrier will always be more effective at high frequencies than at lower frequencies. The pattern of sound radiation from a noise source (e.g., a loudspeaker) will depend very much on its size relative to the sound wavelength. For a loudspeaker, the sound pattern that it radiates will become more directional with increasing frequency. The same is true for the noise radiated from machinery and other noise sources.

These characteristics have practical implications for the noise-control engineer, who can sometimes reduce the effect of high-frequency noise by shielding, or by using the source direction to direct the sound away from the receiver. The resonance frequencies of lengths of pipes or ducts, and of certain types of reactive silencers, depend on the wavelength of the fluid (liquid or gas) inside the pipe or duct, which will in turn depend on the temperature of the fluid inside the pipe or duct. Therefore, the noise-control engineer designing silencers of this type for hot flue gases, for example, will need to take into account how the speed of sound in the gas varies with temperature because this will affect the wavelength of sound in the air or gas travelling through them.

1.3 The Decibel Scale

The dB scale is a logarithmic scale, so first a very brief introduction to logarithms is warranted.

If we write 100 as 10×10, or as 10^2, then 2 is called the logarithm of 100 to the base of 10.

On this same basis:

- Log 10 = 1

- Log 100 = log 10^2 = 2

- Log 1000 = log 10^3 = 3

- Log 1,000,000 = log 10^6 = 6,

- Log 1,000,000,000000 (i.e., one million million) = log 10^{12} = 12

Hence, we can see that a range of numbers from 1 to 1,000 becomes, on a log scale, compressed to a range from 1 to 3, and a range from 1 to a million (10^6) becomes on a log scale a range from 1 to 6, and a range of million million (= 10^{12}) to a range from 1 to 12.

This is what happens when we express the enormous range of sound energies or sound powers (rate of supplying or of using up energy) on the dB scale.

1.3.1 'Level' Means 'dB'

On the dB scale, sound pressures are called sound pressure levels (the word level indicating that a dB scale is being used). The threshold of hearing (20×10^{-6} Pa) becomes a sound pressure level (or sometimes just a sound level) of 0 dB, and the threshold of pain (100 Pa) becomes 120 dB.

1.3.2 What do Changes in dB Level Mean in Ordinary (Non-Decibel) Terms?

A change in level of 3 dB represents a doubling or halving of sound power or sound energy (power is energy per unit time). Hence, if a single source of noise, an electric drill, for example, produces a sound level of say, 85 dBA, at a particular position, and if a second drill produces exactly the same level at this position, then when the second drill is switched on the combined sound level will increase by 3 to 88 dB.

The increases for different numbers of identical sources are shown in **Table 1.3** below:

Another way of looking at **Table 1.3** is that an increase of 10 dB represents a 10-fold increase in sound power or sound energy (i.e., 10 noise sources all operating together), and 20 dB represents a 100-fold increase, and 30 dB a 1,000-fold increase, and so on. Conversely, if a reduction of 20 dB in noise level is required, this means that the sound power must be reduced to one-hundredth of its present value.

Table 1.3 How the dB level increases for multiple identical noise sources	
Quantity of different sources	Increase in level in dB
2	3
3	5
4	6
5	7
6	8
7	8.5
8	9
9	9.5
10	10
100	20
1,000	30

1.3.3 Subjective Effects of Changes in Level

So far we have been considering objective increases or decreases in level, but it is also important to understand the subjective hearing sensation that these changes in sound level will produce. The hearing system itself has a logarithmic response, which is a second reason for using dB (in addition to its convenience in compressing the enormously wide range of sound powers into a manageable scale).

Subjectively, for most people:

- An increase of 3 dB produces a noticeable increase in loudness.

- An increase of 10 dB results in a doubling of loudness.

- A further increase of 10 dB (i.e., an increase of 20 dB) results if a 4-fold increase in loudness.

1.3.4 Combining Sound Levels

When more than one noise source is operating at once it becomes necessary to consider how the individual sound pressure levels combine. The dB values are based on logarithms, so we should not expect them to obey the rules of 'ordinary' arithmetic. We demonstrated above that a doubling of sound energy, power or intensity corresponds to an increase of 3 dB. Hence, if two machines each individually produce a level of say, 90 dB, at a certain point, then when both are operating together we should expect the combined sound pressure levels to increase to 93 dB, but certainly not to 180 dB!

Table 1.4 gives a simple method for combining levels in pairs based on adding to the higher level a correction that depends upon the difference between the two levels. Although this is only an approximate method, it should give results that are accurate to the nearest dB, which is satisfactory for most purposes.

Table 1.4 Combining decibels	
Add to the higher level	Difference between levels
3	0
3	1
2	2
2	3
1	4
1	5
1	6
1	7
1	8
1	9
0	10
Reproduced with permission from R.J. Peters in *Noise Control (A Pira Environmental Guide)*, Pira, Leatherhead, UK, 2000. ©2000, Pira	

Example:

As an example of the use of **Table 1.4**, consider the combining of four dB levels: 82, 84, 86 and 88.

The levels are combined in pairs using **Table 1.4**. The first two levels in the series, 82 and 84 (a difference of 2), are combined to give 86 dB. This 'running total' is then combined with the next in the list, 86 dB (a difference of 0), to give a new running total of 89 dB, which is then combined with the final value, 88 dB (a difference of 1), to give a total combined level of 92 dB.

According to **Table 1.4**, differences of more than 10 dB are negligible, so the lower of the two levels can be ignored. Although the levels may be taken in any order, it is convenient to take them in ascending order, as in this example, so that lower values can be combined first, and so may make a significant contribution when combined with the higher levels.

1.3.5 Prioritising Control of Multiple Noise Sources

The manners in which levels combine with higher levels that always dominate the combined total have important consequences for prioritising noise control.

If, for example, the noise level in a workshop was produced by 10 noise sources, all producing a noise level of 80 dB at a certain point, then the total combined noise level would be 90 dB. The best strategy for noise reduction in this situation (where there is no one dominant noise source) would be to ty to reduce the noise level from all the sources by the same amount. A modest noise reduction of 3 dB from each source, for example, would reduce the overall level by 3 to 87 dB. In contrast, more extensive (and probably more expensive) treatment to reduce the noise from just 3 dB of the sources by more than 10 dB (or even switching them off altogether) would (according to **Table 1.3**) reduce the overall level by about 1.5 dB.

If, however, an eleventh source is introduced into the workshop which produces a noise level on its own of 90 dB, so that there is now one dominant noise source, then the strategy for noise control must be changed. The total noise level would now be 93 dB, and in this situation the obvious priority would be to try to reduce the noise level from the highest source (90 dB) until it was reduced to 80 dB, and then the priority would be to revert, as before, to try to reduce all eleven sources equally.

Successful noise control is often a multi-stage process carried out systematically and with persistence. Also, after each stage of noise control, it is important to re-evaluate the priorities for noise control, as illustrated by the simple example given above.

1.3.5.1 Segregating Quiet and Noisy Areas

Consider a situation in which a new noisy machine which produces a noise level of about 90 dBA is to be introduced into a workplace where there is a quiet area where the noise level is about 70 dBA and a noisy area where the noise level is about 90 dBA. If the new machine is sited in the already noisy area, it will cause the noise level to increase by about 3 dB (i.e., from 90 to 93 dB) but, in the quiet area, the level remains unchanged. Alternatively, if the new noise source is introduced into the quieter area, the noise level there will increase from 70 to 90 dBA, and there will now be two noisy areas to deal with, rather than just one as before.

Therefore, it is best (wherever possible) to collect noisy sources and processes together and to separate them in a different room from quieter sources and activities, and, as far as possible to arrange for as many people as possible to work in the quiet areas and as few as possible in the noisy areas.

1.3.6 Average Sound Levels

Sound levels, whether in the workplace or in the environment, usually vary with time. In a noisy workplace, for example, noise levels might sometimes exceed 90 dBA or even more, but they may also drop as low as 60 or 70 dBA during quieter periods, and similarly wide variations can occur, for example, outside a house near to a roadside. Therefore, in such situations an average noise level is measured. This is also called the equivalent (or average) sound (or noise) level in dB over a period of time (LAeq,T). It is defined as the continuous level of sound (or noise) which, over the measurement period, contains the same amount of A-weighted sound energy as the actual time varying noise being measured.

The fact that the average level is defined in terms of the equivalent amount of energy has many important consequences for the control of noise levels and for hearing protection in the workplace. According to this 'equal energy principle', a reduction in level of 3 dB is equivalent, in sound-energy terms, to a halving of the noise-exposure time. On this basis, therefore, 85 dBA for 8 h is equivalent to 88 for 4 h, which is equivalent to:

- 91 dBA for 2 h

- 94 dBA for 2 h

- 97 for dBA 1 h

- 100 dBA for 30 min

- 103 dBA for 15 min

It can be seen that high levels of noise, even for short periods of time, can lead to high levels of noise exposure. As a consequence, it is always important to prioritise action to reduce the highest noise levels in the workplace, which may mean focussing attention on one dominant noise source rather than spreading effort over all equally. A second consequence relates to the wearing of hearing protectors. Hearing protectors selected and fitted appropriately can provide adequate protection for very high levels of noise. If there is a very high level of noise, for example, more than 100 dB, then the equal energy principle shows that it is extremely important that they are worn for 100% of the time the workers are exposed to these high levels because otherwise a potentially hearing-damaging noise dose can be received in just the few minutes when they are not being worn.

If there are several sources of noise operating together (e.g., several items of machinery), it is always the higher level that will dominate the total combined levels or the average level. For example, as we have already seen, if we combine two sound levels each of 80 dB, then the result will be total of 83 dB but, for two levels of 80 and 70 dB, the combined total is just under 80.5 dB (i.e., the second lower level contributes less than 0.5 dB to the total).

In the same way, higher levels will dominate average or LAeq,T values. Suppose, for example, an employee works for 4 h in a noise level of 60 dB and for the 4 h in a noise level of 90 dB. If the ordinary rules of arithmetic are applied, one might expect the average level over the 8 h working period to be 75 dB, but this is not the case because the 20 dB difference represents a 100-fold increase in sound power and the total amount of sound energy received over the 8 h will, therefore, be almost entirely due to the 4 h exposure at the higher level. Taking this into account, the LAeq,T value over 8 h will be just over 87 dB.

This is an extreme example but it is always true that the true dB average, sometimes called the logarithmic or 'log' average, will always be higher than the ordinary arithmetic average of the levels.

Average sound level (LAeq,T values) are often measured over periods from a few minutes to several hours, although it is also possible to measure sound levels over shorter periods (e.g., less than 1 s).

1.4 Maximum and Peak Sound Pressure Levels

In addition to measuring average, or LAeq,T values, over periods of time, it is sometimes useful to be able to measure the variations of sound level from moment to moment. This is done, for example, to observe the noise levels produced by individual events, such as different machines being operated in a workplace, or noise from shouting, car doors slamming or trains passing, on the environmental noise at a site.

For the measurement of such moment-to-moment or instantaneous sound levels, there are two options: time-weighted sound levels or peak sound levels.

1.4.1 Time-Weighted Sound Levels

Measurement of time-weighted sound levels displays the continuous variations in the running exponentially time-weighted sound pressure levels using fast-time weighting (F) or slow-time weighting (S). The time weightings for these two weightings are 0.125 s (F) and 1.0 s (S). The highest level recorded during a time-weighted sound-level measurement is called the 'maximum sound level(s)' (LAmax), and it is important to state which time weighting (F or S) is used.

In addition to the F and S options, the sound-level meter may also have the option to measure and record peak sound levels.

1.4.1.1 Peak Sound Levels

Impulsive sounds, such as handclaps, hammering, riveting, nail guns, percussive drills, punch presses, and gunfire, can have durations as short as only a few milliseconds. These very rapid fluctuations will be 'smoothed out' by the averaging carried out by the F or S time weightings. If it is required to measure the highest sound levels attained during short bursts of sound such as those mentioned above, then it is necessary to use a sound-level meter that has a peak detection capability, and can measure the peak sound pressure (in Pa) or peak sound pressure levels (in dB).

Time-weighted averaging (using F or S) can be used in a preliminary noise-exposure survey to indicate if continuously varying noise levels at workers' ear positions can sometimes reach 80 or 85 dBA when certain items of machinery are being used. If they do, then this indicates that a more detailed assessment of noise-exposure level must be carried out in which LAeq,T values are measured over appropriate exposure durations.

When measuring impulsive or impact noise from noise sources in the workplace, to assess risk to hearing damage and compliance with *Noise at Work Regulations*, it is the 'peak' noise levels which must be measured and recorded.

If maximum and peak sound levels are measured simultaneously for the same impulsive type of noise event, then the peak sound level will always be higher than the LAmax.

In summary, when measuring noise levels in the environment and wishing to know the highest instantaneous noise levels from short-duration bursts of high-level noise which might cause annoyance to neighbours late at night [e.g., from people shouting, from the slamming of doors (car or lorry), or from bangs/clatters from loading/ unloading of vehicles late at night], it is the 'maximum' noise level, using the F or

S time weighting, which is relevant and must be used. When measuring noise in the workplace for the purpose of assessment of hearing damage, risk assessment, or compliance with *Noise at Work Regulations*, the 'peak' sound level must be used.

1.5 Sound Pressure Levels and Sound Power Levels

Sound pressure and sound pressure level refers to the magnitude of sound-pressure fluctuation produced by one or more sources of sound, and experienced by the human ear or measured by a microphone at a particular position. Sound power and sound power level refer to the total amount of sound power emitted by a sound source.

If we want to describe or specify the amount of sound energy emitted from a sound source, such as an item of machinery, it would be possible to do so in terms of the sound pressure level that it creates at a particular distance from the source, or alternatively in terms of its sound power level. The first (sound pressure level) method of specifying sound emission from sources also requires details of the acoustic environment surrounding the source to be specified, for example, whether the sound source is located indoors (e.g., the type of room, and the presence of nearby sound-reflecting or -absorbing surfaces) or outdoors.

The preferred method for specifying sound emission is to detail the total amount of sound power generated by the sound source (i.e., the rate at which the source emits sound energy). The sound power of the source is measured in watts (W). Like sound pressure, it is common practice to express the sound power levels of sound sources on a dB scale (a different dB scale than that used for sound pressures) and is commonly denoted as Lw.

Sound sources have a tremendous range of sound powers, varying from about 10^{-9} W for the human voice when whispering, to millions of W radiated by a space rocket during launching. The human voice radiates about 20×10^{-6} W during conversation, and this can increase to about 10^{-3} W when shouting. A pneumatic drill used for road-breaking radiates about 1 W, and a typical figure for the noise output of a jet airliner is about 50,000 W.

The sound power scale is based on a reference level of 0 dB representing a source sound power of 1×10^{-12} W (one-millionth of one-millionth of a W). On that basis, a sound power of 1 W corresponds to a sound power level of 120 dB. Many items of industrial and construction equipment have sound power levels in the range 100–120 dB. The conversational sound power level of 20×10^{-6} W corresponds to a sound power level of about 73 dB and that of a shouting voice (1×10^{-3} W) to a sound power level of 90 dB (i.e., 1×10^{-12} W).

Sound power levels, just like sound pressure levels, can be specified and measured either in octave-band levels, or in dBA, dBC or dBZ.

1.5.1 Relationship between Sound Pressure Levels and Sound Power Levels

An approximate indication of the relationship between the sound power level and the sound pressure level produced by small sound source, such a fan hair dryer or electric drill operating well away from any reflecting surfaces and 1 m from the source, is given in **Equation 1.2**:

$$L_p = \text{sound power level} - 11 \text{ dB} \tag{1.2}$$

Hence, for example, for a source of sound power level of 100 dB, the sound pressure level at 1 m from the source would be 89 dB. The sound level would then decrease at a rate of 6 dB for every doubling of distance. This calculation should be used only as an approximate guide because other factors, such as reflection from nearby surfaces and the size of the source, can influence the sound pressure level.

1.6 Sound-Generating Mechanisms: Noise from Vibrating Surfaces, from Air Flow and from Impacts

If we are to try to make things quieter, than we first need to understand how sound is generated, where it come from, and how it arises. We need to know why air molecules suddenly start to vibrate in a particular way – because it is this vibration that reaches our eardrums and causes them to vibrate in a similar way – which cause the sensation of sound in our ears.

There are three ways in which sound and, therefore, noise, is generated: from fluid flow (aerodynamic or hydrodynamic noise) from vibrating surfaces, and from impacts between solid objects.

In the first way, a disturbance to the air causes the air particles to vibrate to and fro. The disturbance can be because the air is moving (called 'flow-generated noise') or because a solid object is moving through the air, disturbing it and causing it to vibrate. This type of noise is called 'aerodynamic noise', or, if it occurs in water, 'hydrodynamic noise', and can be caused by fans, pumps and jets such as exhaust nozzles, by use of compressed airlines, from release of compressed air by pneumatic devices, and by valves

The second way is because the air is set into vibration by a nearby vibrating surface. The layer of air next to the surface is caused to vibrate (in the same way that our eardrums vibrate when sound waves enter the ear canal). Because air is an elastic medium (i.e., it stretches and compresses), the vibration of the layer of air next to the sound source (the vibrating surface) transmits the vibration to the adjacent layer of air, which also vibrates, and so on. Hence, the sound is transmitted by the layer of air next to the source to the layer of air next to the ears.

The third method is noise from impacts (e.g., the noise that occurs when two moving objects collide) or an object is dropped onto a hard floor.

Having introduced these three ways in which sound and noise are generated, it is useful to examine each in more detail to see if there are fundamental clues to ways in which noise may be reduced.

1.6.1 Air Flow-Generated Noise – Aerodynamic Noise

When air flows down a tube or pipe we can distinguish between two types or conditions of flow. If the flow speed is slow and the pipe walls are smooth, the flow will be streamlined or laminar. However, if the flow speed is increased or the tube walls are rough then, at a certain flow speed, the laminar flow breaks up and the flow becomes turbulent. Streamlined, or laminar flow, is fairly quiet, but turbulent flow excites the air into vibration and is very noisy.

The reason for this phenomenon is that when air flows down a tube or pipe, there is a slight difference in flow speeds between different layers so that the air next to the walls of the pipe moves more slowly than the air layer in the centre of the pipe. There is then a drag or frictional force between the layers of adjacent air moving at slightly different speeds. These drag forces give rise to the property of the fluid called 'viscosity', which determines how easily a fluid will flow. When the flow is streamlined, the drag forces become large enough to cause the breakdown of streamlined flow and cause turbulence.

The mathematical theory which describes aerodynamic sound generation is very complicated but, fortunately, it provides two very simple principles that can be used to minimise and reduce flow-generated noise. The first is that to minimise noise, the flow velocity should be kept as slow and as smooth as possible. That is, there should be a minimum number of bends, branches and changes of cross-section in an air- or water-flow delivery system.

This theory also tells us that the level of flow-generated noise rises dramatically with increasing flow speed, typically increasing by 16 dB per doubling of flow speed. Conversely, of course, this means that the noise level will be reduced by the same amount if the flow speed is halved. Hence, the clear message is that, wherever possible, flow speeds should be reduced to a minimum and that even small reductions in flow speed will produce a worthwhile reduction in flow-generated noise.

Hence, in summary, to minimise air flow-generated noise, the airflow must be as slow and smooth as possible to minimise turbulence, and fans and pumps should be operated at the lowest possible speeds.

1.6.2 Noise from Vibrating Surfaces

Most machinery noise arises from the vibration of the outer surfaces of machines, such as fans, gearboxes, motors and engines. These surfaces are often made of thin sheet steel, and act rather like loudspeakers by turning the vibration into unwanted sound or noise.

So where does the vibration, which causes the noise, come from? It comes from the forces generated by the machine as part of its function (e.g., forces that lift, rotate, cut, grind and so on). Most of these forces are generated in the core structure of the machine, which is usually strong and rigid, to withstand them. However, a small part of these forces are transmitted to the much thinner and more flexible outer surfaces from where the noise is radiated.

Hence, with this simple model of machine-noise generation in mind, what are the opportunities and strategies for reducing noise?

The possible strategies to reduce sound radiation from vibrating surfaces involve trying to: minimise the magnitude of the vibratory forces which are causing the surface to vibrate; isolating theses forces from the vibrating surface; minimising the area of the vibrating surfaces; covering the vibrating surface (i.e., with the use of acoustic lagging or cladding). The various possibilities are discussed again in **Chapter 2** in connection with noise from machinery.

1.6.3 Noise from Impacts

When a hard metal object falls onto a sheet of mild steel, there are two components to the noise that is generated: (i) the impact itself and (ii) the ringing of the impacting components which continues after the impact has ceased. The amount of sound energy generated by the impact depends on the momentum of the impacting parts (i.e., upon their mass and their relative velocity at impact).

Thus, the possibilities for noise reduction are to reduce the: mass of the colliding components or to slow down the collision speed by cushioning the impact with a resilient layer; the drop height. In addition, the impact causes the sheet to 'ring' like a bell. This ringing may be reduced by increasing the damping of the panel.

1.7 Airborne and Structure-borne Sound

Noise from a machine such as a compressor fan or a motor can reach the ears of the receiver by a combination of airborne and structure-borne sound transmission. Airborne sound is radiated directly from the vibrating surfaces of the machine into the surrounding air. In addition, the vibration of the machine may be transmitted to nearby structurally connected surfaces such as walls or floors, pipework or ductwork or to the surface of other connected items of equipment (e.g., from a motor to a fan, or from an engine to a gearbox).

In an indoor situation, airborne sound will be confined to the indoor space in which it is generated, whereas structure-borne sound will be transmitted *via* adjacent walls and floors to adjacent spaces. Each wall or floor will provide a significant reduction in the sound level transmitted to the adjacent spaces whereas, by contrast, sound is often transmitted *via* a structure very efficiently, often with minimum loss.

In addition, 'structure-borne transmission' means that sound is being radiated by a greatly increased surface area than the original source vibrating source area, and in effect amplifies the noise. A good example is the sounding-board effect exploited in the design of many stringed musical instruments and of toy xylophone-based musical boxes.

Therefore, it is particularly important to isolate noise-generating machinery from nearby surfaces that can radiate additional noise.

1.8 Sound Insulation, Sound Absorption and Vibration Isolation and Damping

Three methods can be used to prevent sound being transmitted from source to receiver. These involve the use of sound-insulating and sound-absorbing materials to control airborne sound transmission, and the use of vibration isolation and damping to reduce the effect of structure-borne sound and vibration.

Sound-insulating materials provide a barrier to the transmission of airborne sound. Sound-absorbing materials absorb the airborne sound energy incident on their surface and minimise sound reflections. Resilient materials are used to minimise the transmission of vibration, which leads to the generation of structure-borne sound. Well-damped materials are used to minimise the effects of resonant vibration and of transient vibration caused by impacts

1.8.1 Sound Insulation

Good sound-insulating materials are those that can resist sound transmission. The parameter that defines the property of sound insulation is the sound reduction index (R), also called 'transmission loss', in dB. The sound reduction index of a building element essentially describes the fraction or percentage of sound energy transmitted by the element, but converted into dB. For example:

- A building element with sound reduction index = 20 dB will transmit 1/100 of the sound energy incident upon it.

- A building element with sound reduction index = 30 dB will transmit 1/1,000 of the sound energy incident upon it.

- A building element with sound reduction index = 40 dB will transmit 1/10,000 of the sound energy incident upon it.

- A building element with sound reduction index = 50 dB will transmit 1/100,000 of the sound energy incident upon it.

Sound-insulating materials are used as partitions between indoor spaces (walls, floors and ceilings), as noise reducing screens and barriers, and as complete enclosures around machines. In each case, a requirement is that the sound-insulating material shall have a sufficiently large sound reduction index in each frequency band to provide the necessary sound insulation. It is also important that the insulation is complete, with the minimum amount of holes and gaps around the edges. Good seals are important if high sound insulation is to be achieved. The sound reduction index of a single-leaf partition will depend upon the mass, stiffness and damping of the material but, in many cases, the mass is the most important factor.

The mass law of sound insulation is an approximate law that assumes that the sound reduction index of a partition may be predicted entirely from knowledge of the surface density of a partition expressed in kilograms per square metre. It can be used to estimate the sound reduction index value if data from standardised laboratory tests are not available. According to the mass law, the sound reduction index will increase with frequency by 6 dB for every octave and also will increase by 6 dB for every doubling of mass. Although the mass law may be obeyed approximately by many materials, neglecting the influence of damping and stiffness leads to discrepancies at high and low frequencies. At low frequencies, there are panel resonances. At high frequencies above the mass law region, flexural waves in the panel also cause resonances, a phenomenon called the 'coincidence effect', which occurs above a critical frequency. Some typical sound reduction index values for some single-leaf partitions are shown in **Table 1.4**.

Table 1.5 shows that, in general, the value of sound reduction index increases with increasing frequency, even if the mass law rule of a 6 dB increase per octave band is not always achieved. In some cases, there is a reversal in the trend of increasing values of sound reduction index with increasing frequency, such as for glass and chipboard at 2,000 Hz. These are likely to be reductions due to the coincidence effect.

The value of the sound reduction index of partition between two spaces (e.g., between two adjacent rooms) will be a major factor in determining the difference in sound level between the two rooms arising from sound transmission through the partition. However, other factors can also have an effect, such as the area of the partition and the amount of sound absorption in the receiving room.

The sound reduction of building elements such as walls, floors/ceilings, doors and windows is determined by laboratory tests in sound transmission suite (two adjacent reverberant rooms) carried out in accordance with British Standards European Norms International Organization for Standardization, BS EN ISO 10140-2:2010. This test procedure, which is carried out in octave or 1/3 octave-frequency bands, gives the highest possible value. Designers often use slightly lower values in calculations to predict sound transmission under field conditions because of possible differences in installation conditions, less-than-perfect workmanship, and possible flanking transmission paths. Another standard, BS EN ISO 717-1:1997, gives a method for converting a series of measured 1/3 octave-band values of sound reduction index into one single-figure value: the weighted sound reduction index. This single-value figure is sometimes used in the first, approximate stage of design and specification, but full data for frequency bands may often be needed for the final stages, particularly in critical situations.

Table 1.5 Some typical sound reduction index values of panels and partitions						
Panel/partition type	Octave band (Hz)					
	125	250	500	1 kHz	2 kHz	4 kHz
	Single-leaf panels					
1.5 mm lead sheet	28	32	33	32	32	33
3.0 mm lead sheet	30	31	27	38	44	33
1 mm sheet steel	8	14	20	26	32	38
1.6 mm sheet steel	14	21	27	32	37	43
Plastered brick	36	36	40	46	54	57
Thermalite blocks	27	31	39	45	53	38
Plasterboard	15	20	24	29	32	35
Plywood	9	13	16	21	27	29
Chipboard	17	18	25	30	26	32
6 mm glass	24	26	31	34	30	37
	Double-leaf partitions					
Cavity brick wall (50 mm cavity)	37	41	48	60	61	61
Double glazing (6 mm glass, 12 mm cavity, 6 mm glass)	26	23	32	38	37	52
Plasterboard stud partition (two 12.5 mm plasterboard sheets with 50 mm cavity filled with glass fibre quilt)	21	35	45	47	47	43
Machine enclosure panels (16 gauge steel, 100 mm mineral fibre, 22 gauge perforated steel sheet)	21	27	38	48	58	67

Reproduced with permission from R.J. Peters in *Noise Control (A Pira Environmental Guide)*, Pira, Leatherhead, UK, 2000. ©2000, Pira

1.8.2 Double-Leaf Partitions

Although the sound reduction index of a partition may be increased by increasing its thickness and thereby its mass, an increase of only 6 dB is achieved even for a doubling of mass and thickness. Hence, trying to increase insulation by this means rapidly leads to a law of diminishing returns. One way of getting around this limitation is to use double-leaf constructions in which two sound-insulating 'leaves' are separated by a cavity. There are, however, some important design principles if double-leaf constructions are to be effective:

- The cavity should be filled as far as possible with a sound-absorbing material to reduce the build-up of reverberant sound within the cavity.

- Flanking of the double-leaf structure as a result of structure-borne transmission *via* the framework between the two leaves must be prevented. This can be achieved by isolating each leaf from the framework and from the other leaf.

These design features are illustrated in **Figure 1.3**.

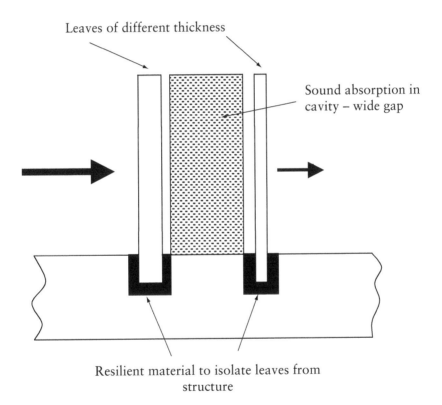

Figure 1.3 Design features of a double-leaf construction necessary to achieve good sound insulation. Reproduced with permission from R.J. Peters in *Noise Control (A Pira Environmental Guide)*, Pira, Leatherhead, UK, 2000. ©2000, Pira

1.8.3 Sound Absorption

Sound-absorbing materials reduce noise by reducing the reflection of sound at surfaces and turning sound energy into heat *via* frictional processes within the material. Sound-absorbing materials may be used in rooms and other spaces to reduce reverberant sound, to line ducts and, as indicated above, to absorb sound between the leaves of double-skinned structures.

The parameter that defines the property of sound insulation is the sound-absorption coefficient (α), which is a quantity with a value between 0 and 1. It is the fraction of the sound energy that is incident upon a surface which is not reflected or transmitted. Hence, a value of $\alpha = 1$ represents a perfect sound absorber and a value of $\alpha = 0$ represents a perfect reflector of sound.

The α of materials is measured by one of two standard methods, as defined in BS EN ISO 354:2003 and BS EN ISO 10534-1:2001. These methods allow the α to be measured at different frequencies, usually in octave or 1/3 octave bands. Another standard procedure, as defined in BS EN ISO 11654:1997, provides a method for converting these frequency-related data into a single figure rating value (class A, B or C) for sound absorption, which is often used in the manufacturer's promotional literature, but octave-band data should be obtained and used wherever possible.

The type of sound absorber most likely to be used in industrial situations is the porous absorber. In a porous absorber, sound energy is converted to heat *via* the friction that occurs inside the material between the vibrating air particles and absorber material (usually foam or fibrous material). This process will give some sound absorption at all frequencies (which is why they are sometimes called 'broadband absorbers') but the amount of absorption will be poor at low frequencies but increase with frequency, and there can be near-perfect absorption at high frequencies. The amount of absorption also depends on the thickness of the sound-absorbing layer, with thicker layers giving more sound absorption.

Table 1.6 shows some typical values of the α for some common materials. It can be seen that hard surfaces such as steel, concrete, brick or plaster and plasterboard have a very low α. Hence, in internal spaces in which the surfaces are predominantly made of these hard, sound-reflecting materials, there will be multiple sound reflection before the sound energy is eventually absorbed by the room surfaces, and the sound will behave rather like a hard steel ball bouncing many times off a hard floor. Such a room will be acoustically 'live' or 'reverberant'. In contrast, a room with carpets, an acoustically tiled ceiling and curtains and soft furnishing (all with a high α), there will be very little sound reflection, and the room will be acoustically 'dead'. Production spaces and workshops containing machinery are often acoustically 'live' spaces with a high level of reverberant sound.

1.8.3.1 Reduction of Reverberant Sound Levels by Increasing Sound Absorption

Additional sound absorption introduced into a room or space will reduce the level of reverberant noise in a workplace (i.e., the noise that reaches the receiver after being reflected or scattered from the room surfaces, as opposed to travelling directly from the noise source to the receiver). Additional sound-absorbing materials may be used to line the walls, floor and ceiling of the space or, if this is not practicable, may be hung as 'space absorbers' from the ceiling. The amount of sound reduction will depend on the degree to which the amount of sound absorption in the room is increased. Hence, this treatment will be most effective for rooms and spaces that have initially very little absorption in them (i.e., rooms with hard surfaces, such as plaster or concrete). This is the case with many factory spaces. The level of reverberant sound decreases by 3 dB each time the amount of absorption is doubled (i.e., a reduction of 6 dB for a 4-fold increase in the amount of absorption, or a reduction of 10 dB if the amount of absorption is increased 10-fold).

Even hard surfaces such as brickwork, plaster and concrete provide some absorption, so the amount by which reverberant sound may be reduced by introducing extra absorption is limited. For example, a typical α value for these materials would be between 0.01 and 0.05 at low and medium frequencies, respectively. To take an extreme example, if an engineering workshop, with all its surfaces being hard and sound reflecting, were to be re-fitted as a well-furnished office with a carpeted floor and sound-absorbing ceiling, this might result in an approximate 10-fold increase in the average α, from about 0.05 to about 0.5, and this would produce a 10 dB reduction in the reverberant sound level. Even theoretically there is not the possibility of another 10-fold increase to increase the average α to 1. Therefore, this method of reducing noise can never produce a reduction of more than 20 dB, and in practice the limit is usually less than 10 dB and often only around more than 5 dB. Nevertheless, the subjective response to such a reduction is often favourable because the reduction is entirely of reverberant sound and the effect may be to produce a less diffuse sound field in the space so that the direction from which the sound is coming (and hence the source of the noise) is more apparent. The noise-control treatment is also very apparent visually.

This type of sound-absorbing treatment will not provide any noise-reduction benefit for a person who works close to a noisy machine and receives the major part of his/her noise dose directly from that source, and not from reverberant sound.

Material type	Octave band					
	125 Hz	250 Hz	500 Hz	1 kHz	2 kHz	4 kHz
Brick	0.05	0.04	0.02	0.04	0.05	0.05
Concrete	0.01	0.01	0.02	0.02	0.02	0.02
Plaster (on brick or block)	0.04	0.05	0.06	0.08	0.04	0.06
Plasterboard	0.3	0.2	0.15	0.05	0.05	0.05
Steel	0.05	0.05	0.05	0.05	0.05	0.05
Glass	0.25	0.25	0.18	0.12	0.07	0.05
Timber	0.15	0.11	0.1	0.07	0.09	0.03
Carpet (thick pile)	0.15	0.25	0.5	0.6	0.7	0.7
Carpet (standard)	0.02	0.06	0.15	0.4	0.6	0.7
Curtains (heavy)	0.07	0.3	0.5	0.75	0.7	0.6
Mineral fibre (25 mm)	0.07	0.25	0.5	0.8	0.9	0.8
Mineral fibre (50 mm)	0.39	0.91	0.99	0.97	0.94	0.89

Table 1.6 Typical values of sound-absorption coefficient for some common materials

Reproduced with permission from R.J. Peters in *Noise Control (A Pira Environmental Guide)*, Pira, Leatherhead, UK, 2000. ©2000, Pira

1.8.4 Vibration Isolation

Vibration isolation is the reduction of the transmission of vibration or structure-borne sound in a machine or building by the use of resilient material placed between the source and the structure of the machine or building. The simplest example is a machine such as a pump or motor placed on springs or anti-vibration mounts to reduce the vibration from the machine to the floor beneath (**Figure 1.4**).

The important property of such a system is called the 'transmissibility', and this is the ratio of the vibration amplitude at the base of the machine to that on the floor beneath (i.e., the ratio of the vibration amplitude across the isolator or spring). **Figure 1.4** shows how this transmissibility varies with the forcing frequency that, in the case of a rotating machine such as a pump or motor, is related to the rotational speed in revolutions per minute (rpm). The machine and isolator system are modelled as a mass and spring system in which the machine acts as the mass on top of the isolator, which is the spring. Such a mass–spring system has its own natural frequency based upon two parameters: mass and stiffness.

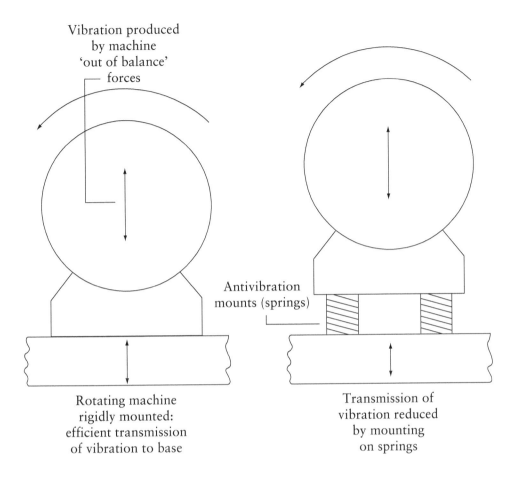

Figure 1.4 The isolation of vibration from a machine using an anti-vibration mount. Reproduced with permission from R.J. Peters in *Noise Control (A Pira Environmental Guide)*, Pira, Leatherhead, UK, 2000. ©2000, Pira

The graph of **Figure 1.5** shows that at low speeds corresponding to forcing frequencies below the natural frequency of the mass–spring system, the transmissibility is close to unity. That is, the springs are in effect rigid and all of the vibration from the motor is transmitted to the floor below. As the forcing frequency (i.e., the speed of the machine) is increased, the transmissibility increases to a sharp peak at the speed corresponding to the natural frequency of the mass–spring system. This is the phenomenon of 'resonance' and in this situation the isolators are making matters worse and the vibration transmitted to the floor will be amplified rather than reduced. As the frequency increases above the resonance, the transmissibility decreases until at a frequency of 1.4 × natural frequency it falls to 1, and for higher frequencies the transmissibility falls to less than 1. In this region and for all higher frequencies where the transmissibility is less than 1, isolation is occurring and the vibration on the floor is less than that on the motor, and the isolators are having the desired effect.

Although detailed designs will be based on a specific isolation target, an approximate rule of thumb is that a value of f/f_0 more than 3 is needed to provide significant vibration isolation.

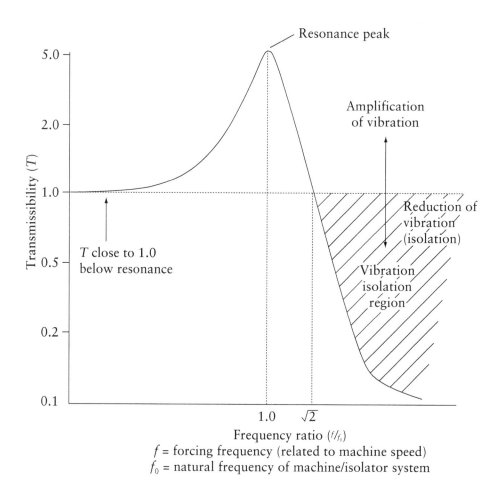

Resonance peak

Amplification of vibration

Reduction of vibration (isolation)

T close to 1.0 below resonance

Vibration isolation region

Frequency ratio (f/f_0)
f = forcing frequency (related to machine speed)
f_0 = natural frequency of machine/isolator system

Figure 1.5 Typical variation of transmissibility with frequency for a machine mounted on vibration isolators. Reproduced with permission from R.J. Peters in *Noise Control (A Pira Environmental Guide)*, Pira, Leatherhead, UK, 2000. ©2000, Pira

Let us suppose that the motor is rotating at 3,000 rpm. This corresponds to $3,000 \div 60 = 50$ revolutions per second, and so results in a vibratory driving force with a driving frequency, f, of 50 Hz. For the mass–spring system to be driven well above resonance then, using the rule of thumb, the natural frequency f_o of the system should be arranged to be no more than $50 \div 3 = 16.7$ Hz.

This is arranged by selecting the stiffness of the springs in conjunction with the mass of the motor. If this is done, the transmitted level of vibration will be much reduced (i.e., the vibration produced by the motor has been isolated, as shown in the right of **Figure 1.5**).

Anti-vibration mounts are normally selected for the lowest usual operating speed of the machine. The stiffness of the spring is chosen to produce a natural frequency when used in conjunction with the mass of the motor, which is well below the frequency corresponding to the normal running speed. Hence, under normal running conditions, the mass–spring system will be forced or driven at a frequency well above its natural frequency, and so the transmissibility will be much less than unity and isolation will be achieved.

Various materials may be used for isolation depending upon the frequency range and machine speed of interest. The natural frequency of the machine/isolator system is also related to the static deflection of the isolators. The 'softer' the isolators (i.e., the lower their spring stiffness) the lower is the frequency for which they will be effective, and the higher the static deflection. Above frequencies of about 25 Hz, corresponding to static deflections of less than 1 mm, sufficient resilience may be achieved using pads of cork or foam plastic or mineral fibre. At lower frequencies from about 15 Hz upwards, corresponding to static deflections of less than 10 mm, vibration isolators made of rubber are common but, for lower frequencies below about 10 or 15 Hz, where static deflections are more than 1 mm, metal springs are usually used. Exactly the same principles used to isolate a complete machine from the structure of a building may be used to isolate parts of a machine from the main framework of the machine, or to isolate whole floors (i.e., floating floors) or ceilings (i.e., suspended ceilings) from the building structure. In all cases, the isolation must be complete and bridging of the isolation or flanking paths must not occur. Therefore, if a pump or motor has been isolated from the floor by anti-vibration mounts, any connection to this machine, such as ducts, pipes, cables or conduits, must be made flexibly, otherwise vibration from the machine could travel down these alternative paths to the floor, thereby short-circuiting or bridging the isolation provided by the anti-vibration mounts.

1.8.5 Materials for Sound Insulation, Sound Absorption and Vibration Isolation

Good sound-insulating materials tend to be solid, heavy and dense, such as sheet metal, brick, concrete, plasterboard, glass, hard plastic and timber. These materials are hard and sound reflecting and, therefore, are poor sound absorbers. Porous sound absorbers are usually porous or open-celled foam-plastic materials which have low density and so are usually poor sound insulators. These two types of material are often used together in noise-control solutions such as sound-reducing barriers, double-leaf partitions and noise enclosures, as discussed in more detail in **Chapter 2**.

The essential feature of material used for isolation is that they are resilient (i.e., they act like springs). This includes metal springs, cork and rubber or other polymer materials. Mineral fibre layers can be used as good sound absorbers, but also as isolating materials (e.g., as the resilient layer in floating floors) and, although they will be poor sound insulators, they are used widely as good thermal insulators. These materials are available with a very wide range of fibre densities, which will affect their performance as sound insulators and vibration isolators.

1.8.6 Damping

'Damping' is a process whereby the energy of vibration in a vibrating structure is converted into heat by some frictional mechanism which thus leads to a reduction in the level of vibration. The presence of damping is important in limiting the vibration and noise generated by transient vibration caused by impacts, and in reducing the level of resonant vibration. It is for this reason, for example, that a 'bottle bank' produces much less noise if it is made of a well-damped plastic material than if it is made of mild steel, which has very low damping, and would cause ringing noise and reverberation each time a bottle is thrown into the bottle bank. Damping is discussed in more detail in **Chapter 2**.

1.9 Summary

The main points covered in this chapter are summarised below:

1. Noise has been defined as 'unwanted sound'. Sounds are caused by minute pressure fluctuations in the air which travel, in the form of waves, from the source to receiver.

2. The human ear can detect a very wide range of sound pressures, with even the most intense sound representing only a very small fluctuation in atmospheric pressure. Because of this phenomenon, the energies involved in noise are minimal even if those sounds produce major effects on listeners. Audible sound pressures range from 20-millionths of a Pa to about 100 Pa, but vary from person to person and also depend upon age.

3. The simplest of all sounds, the pure tone, may be described in terms of its amplitude, frequency (f) and wavelength (λ).

   These last two terms are related by **Equation 1.1** in which 'c' is the velocity of sound in the medium.

4. Frequency is measured in Hz. The audible range of frequencies extends from 20 to 20,000 Hz approximately.

5. The velocity of sound in air varies under normal conditions between 330 and 340 m/s depending on the air temperature. Sound wavelengths vary from a few centimetres at high frequencies (3.3 cm at 10,000 Hz) to several metres at low frequencies (3.3 m at 100 Hz).

6. The frequency content of sounds must be known because almost all acoustic parameters and properties associated with noise measurement and control vary with frequency, including the sensitivity of human hearing, the sound-absorbing and sound-insulating properties of materials, and the noise emission of noise sources.

7. The effectiveness of various noise-control measures, such as attenuators, silencers, acoustic tiles, partitions, enclosures, barriers and anti-vibration mounts, varies with frequency.

   Usually it is easier to achieve noise reductions at high frequencies than at lower frequencies.

8. The frequency of a tone determines the subjective measure of its pitch. The subjective impression created by a noise will also very much depend on its frequency content.

   The most commonly used method of frequency analysis is octave-band analysis, which is appropriate for noise-control purposes. 1/3 octave-band analysis is used for measurements of building acoustics and narrowband analysis is used to identify the frequency and thence the source of pure tones among broadband noise.

9. Wavelength is important in determining how sound travels (i.e., is reflected by surfaces, scattered by obstacles, and the extent to which it bends around barriers). The wavelength of a sound has a profound effect on the way it interacts with obstacles in its path.

10. The dB scale is a logarithmic scale for comparing the ratios of two powers. Sound pressure is related to sound power so it may also be measured on a dB scale.

11. Decibels are used for convenience to compress the enormous range of audible sound pressures into a manageable scale and to simplify sound-transmission calculations not only because multiplying and dividing processes become addition and subtraction processes when using dB, but also because the human response to the stimulus of sound is logarithmic in nature.

12. Because they are based on logarithms, dB levels cannot be added together or averaged like ordinary numbers. The combination of sound pressure levels may be carried out by calculation (handheld calculator) or using special charts or nomograms. Two identical levels will combine to produce an additional 3 dB and ten identical levels will produce an extra 10 dB.

13. Subjectively, an increase of 3 dB is noticeable and an increase of 10 dB represents a doubling of loudness.

14. A knowledge of how sound levels from multiple noise sources combine is necessary to decide the priorities for noise control. The priority must always be to try to reduce the noise emitted from the noisiest source until a situation is reached where all the sources contribute equally to the total noise level, at which point the priority becomes to reduce noise from all sources equally. As far as possible, noisier sources should be separated from quieter sources.

15. dBA is the most commonly used method of measuring noise. It includes the contributions of all the frequencies in a single-figure value to take into account the way in which the sensitivity of human hearing varies with frequency, so that the measured dBA value relates to the average human perception of the magnitude of the noise.

16. The equivalent continuous noise level of a time-varying noise is the constant noise level which, over the period of time under consideration, would contain the same amount of A-weighted sound energy as the time-varying noise.

17. If a range of sound levels are averaged, the decibel or logarithmic method of averaging always produces a higher average value which is much more influenced by the higher levels in the range than is the case with arithmetic averaging.

18. According to the equal energy principle, a reduction in level of 3 dB is equivalent, in sound energy terms, to a halving of the noise-exposure time. On this basis, high levels of noise, even for short periods of time, can lead to high levels of noise exposure. One consequence is that a failure to wear hearing protectors, even for a short duration of exposure, can lead to a significant risk of noise-induced hearing loss.

Peak and Maximum Sound Pressure Levels

19. Peak sound levels must always be used when measuring high levels of impulsive noise in the workplace to evaluate the risk of hearing damage. Peak sound levels are the highest values of instantaneous sound level occurring during the measurement.

    The LAmax is the highest sound level recorded using F or S time weightings. Peak levels will always be higher than maximum levels. The LAmax is used when measuring noise in the environment, or for preliminary checks on noise levels in the workplace, prior to a more thorough assessment of LAeq,T values.

Sound Pressure Levels and Sound Power Levels

20. Sound pressure level refers to the magnitude of sound-pressure fluctuation produced by one or more sources of sound, and experienced by the human ear or measured by a microphone at a particular position. Sound power levels refer to the total amount of sound power emitted by a sound source. Both are measured on a dB scale, but sound pressure level relates to sound pressures measured in Pa whereas sound power levels relate to sound powers measured in W.

21. All noise is generated by fluid flow aerodynamic or hydrodynamic forces, vibrating surfaces, or by impacts. In each of these three cases, some simple general principles can be formulated which can be used to minimise noise production.

22. To minimise air-flow generated noise, the airflow must be as slow and smooth as possible to minimise turbulence, and fans and pumps should be operated at their lowest possible speeds.

23. The possible strategies to reduce sound radiation from vibrating surfaces are: trying to minimise the magnitude of the vibratory forces causing the surface to vibrate; isolating theses forces from the vibrating surface; minimising the area of the vibrating surfaces; covering the vibrating surface (i.e., the use of acoustic lagging or cladding).

24. The possibilities for noise reduction from impacts are to reduce the mass of the colliding components or to slow down the collision speed by cushioning the impact with a resilient layer; or by reducing drop height. In addition, the impact causes the sheet to ring like a bell. The ringing may be reduced by increasing the damping of the panel.

25. Sound (or noise) may travel from the source to the receiver by several paths involving airborne or structure-borne sound transmission.

26. The four principal methods of reducing noise transmission between the source and the receiver are using sound insulation and sound absorption (for airborne sound) and vibration isolation and damping (for structure-borne sound).

27. Sound-insulating materials are solid, dense materials used as partitions to minimise sound transmission. The sound reduction index, measured in dB, is the property which expresses the degree of sound insulation provided by a building element such as a wall, floor, door or window. Although there are exceptions, sound reduction index values tend to increase with frequency.

28. Sound-absorbing materials are used to minimise sound reflections by using the frictional processes occurring within the material to convert sound energy into heat. They are used to reduce reverberant sound levels in indoor spaces, screens, double-leaf partitions and in noise-reducing enclosures. The most commonly used sound absorbers are porous absorbers. They are more effective at medium and high frequencies than at low frequencies.

29. Isolating materials are used to reduce the transmission of vibration and structure-borne noise. Materials used for isolation are elastic and resilient materials such as cork, mineral fibre or plastic foam pads (for light loads), and rubber pads used in shear configuration or metal springs for heavier loads. Structure-borne sound travels very efficiently though structures, and use of isolating material to prevent the spread of structure-borne sound is a very important aspect of vibration isolation.

30. Well-damped chutes, hoppers, and conveyor belts will minimise impact-generated noise and will also reduce the effects of resonance.

Bibliography

1. I.J. Sharland in *Woods Practical Guide to Noise Control*, Woods of Colchester Ltd., (Flakt Woods), Colchester, UK, 1972.

2. *Noise Control in Industry*, Ed., J.D. Webb, Sound Research Laboratories Ltd., Altrincham, UK, 1976.

3. *Noise Control in Mechanical Services*, Sound Research Laboratories Ltd., Altrincham, UK, 1976

4. R.J. Peters, B.J. Smith and M. Hollins in *Acoustics and Noise Control*, 3rd Edition, Prentice Hall, Upper Saddle River, NJ, USA, 2011.

5. D.A. Bies and C.H. Hansen in *Engineering Noise Control*, 2nd Edition, E & FN Spon, London, UK, 1996.

6. R.J. Peters in *Noise Control (A Pira Environmental Guide)*, Pira, Leatherhead, UK, 2000.

7. BS EN ISO 10140-2:2010 – Acoustics – Laboratory measurement of sound insulation of building elements – Measurement of airborne sound insulation.

8. BS EN ISO 717-1:1997 – Acoustics – Rating of sound insulation in buildings and of building elements – Airborne sound insulation.

9. BS EN ISO 354:2003 – Acoustics – Measurement of sound absorption in a reverberation room.

10. BS EN ISO 10534-1:2001 – Acoustics – Determination of sound absorption coefficient and impedance in impedances tubes – Method using standing wave ratio.

11. BS EN ISO 11654:1997 – Acoustics – Sound absorbers for use in buildings – Rating of sound absorption.

2 Principles of Noise Generation and Control

2.1 Introduction

The aim of this chapter is to review and explain the range of methods available for reducing noise, starting with common sense 'good housekeeping' methods. The use of engineering methods to reduce noise involves understanding how noise is produced, and so the noise-generating mechanisms involved in some common sources of noise in industry are explained, and the principles of noise reduction described. The reduction of noise during its transmission path between the source and receiver, using the methods of sound insulation, sound absorption and vibration isolation, are discussed, as well as methods for the control of machinery noise at source. Methods for diagnosing the most important noise sources and paths and for the specification of noise from machinery are described.

2.2 Good Planning, Management and Housekeeping

Although there are many technical solutions to noise problems, the best approach is to avoid or minimise problems wherever possible so that such solutions are not necessary:

- Avoid using noisy machines and processes if quieter alternatives are available.

- Avoid making unnecessary noise, for example, leaving machines running when not necessary.

- Locate noisy machines and processes away from noise-sensitive areas.

- Avoid making excessive noise at sensitive times, for example, late at night or at weekends.

- Consider noise issues when planning changes to operations in the future so that steps can be taken well in advance to minimise noise exposure and disturbance.

- Maintain machinery and equipment in good repair so that unnecessary increases in noise levels are avoided. This also applies to the fabric of buildings so that doors and windows, for example, are always a good fit and do not transmit noise through holes or gaps.

2.3 Mechanisms of Noise Generation

It is useful to identify three main mechanisms: the radiation of sound from a vibrating solid surface to the adjacent fluid, the direct disturbance of the fluid flow, and the noise generated by impacts.

The first category includes noise from machine panels and the external casings of motors, gearboxes and fans. Many machines have a stiff inner framework designed to withstand the working forces generated by the machine, such as cuttings, drillings and pressings, but the outer covers are often of thin sheet steel panels which vibrate and radiate noise.

The second category includes all 'aerodynamic' and 'hydrodynamic' noise, in which sound is generated by a disturbance to the fluid that causes turbulence. This includes noise from fans, pumps, whistles and jets.

The third category of impact-generated noise includes not only machinery based on impacting processes such as riveting, punching and pressing, but also mechanical handling processes where items fall into trays, hoppers and bins.

In each case it is possible to identify the main factors that affect noise radiation and which, therefore, provide some basic principles for noise control. The sound power radiated from a vibrating panel will depend on the vibration amplitude of the panels, its area, and on a factor called its radiation efficiency. A reduction in any of these will reduce the sound energy radiated. The possibilities for reducing the vibration amplitude are to reduce the vibratory forces reaching the panel from the machine frame, and then to design the panel so that these forces produce the minimum vibration amplitude. The first aim may be achieved by reducing the forces themselves, at source, or by preventing their transmission to the panel. The mass, stiffness or amount of damping in the panel will affect its vibration response. If the noise-radiating panel is a machinery guard, it may be possible to use a cover in the form of a grille rather than a solid sheet to reduce the noise-radiating area. The radiation efficiency of a panel depends upon its size compared with the wavelength of the sound, and upon the panel stiffness.

Noise generated by fluid flow increases rapidly with flow velocity, and so an important principle is to keep the flow speed as low as possible. Turbulent flow conditions produce much more noise than streamlined (laminar) flow, and so any factors which cause turbulence, such as sudden changes in flow conditions, changes in flow speed/direction or obstacles in the fluid which spoil smooth flow, will cause extra noise. Therefore, another important principle is to ensure that flow conditions are always as smooth as possible.

If a hard metal object falls onto a sheet of mild steel there are two components to the noise that is generated: the impact itself, and the 'ringing' of the impacting components which continues after the impact has ceased. The amount of sound energy generated by the impact depends on the momentum of the impacting parts (i.e., upon their mass and their relative velocity) at impact.

Thus, the possibilities for noise reduction are to reduce the: mass of the colliding components or to slow down the collision speed by cushioning the impact with a resilient layer; or to reduce drop height. In addition, the impact causes the sheet to ring like a bell. The ringing may be reduced by increasing the damping of the panel.

2.4 The 'Source Path Receive' Model of Noise Control

The simplest model of sound transmission shows the noise travelling from the source *via* a transmission path to a receiver (**Figure 2.1a**). Even with this simple model there are three ways to reduce the noise level: (i) at source; (ii) by the transmission path; and (iii) at the receiver.

An example would be when an individual sitting in his garden is affected by sound from a radio in a neighbour's garden, where the path is direct transmission through the air. Control at source would involve the neighbour reducing the volume. Control *via* transmission might be achieved by moving away (i.e., increasing the distance between source and receiver) or by erecting some sort of barrier. Control at the receiver might involve the receiver wearing ear plugs or moving indoors.

It is often stated that noise control at source is the preferred option. The reason for this is best understood by considering an extension to the simple model in which sound is transmitted to a single receiver by several different paths (**Figure 2.1b**). Control during the transmission stage would now require three control remedies, one for each path, whereas a single noise-control measure at source would reduce the transmission *via* each of the paths.

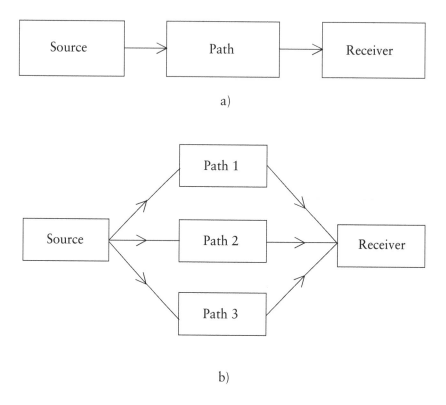

a)

b)

Figure 2.1 Simple source-path receiver models of sound transmission

A very simple example is illustrated in **Figure 2.2**, in which the noise source is a machine in a room. Sound from the source reaches the receiver, in the same room, by three paths. The first of these is direct airborne sound. The second is also airborne sound but reaches the receiver after multiple reflections by surfaces in the room. This is often called the 'reverberant' component of the airborne sound. The third transmission path, called 'structure-borne sound', arises from transmission of vibration energy from the source into the building structure (*via* the floor in this case) and subsequent radiation from the structure as airborne sound.

The methods for controlling the sound transmission *via* these three paths use the standard methods of sound insulation, sound absorption, and vibration isolation (**Figure 2.2**). Transmission *via* the first path may be reduced by interposing a sound-insulating screen between the source and receiver. Although this would reduce the transmission of the direct airborne sound, it would not be completely effective on its own because sound would be able to 'bypass' the screen and reach the receiver *via* reflections at room surfaces. This path may be reduced by lining the room surfaces with sound-absorbing materials, which would reduce the amount of reverberant sound in the room. However, structure-borne sound may still reach the receiver unless the third path is controlled by the use of a vibration-isolating material placed between the source and the structure.

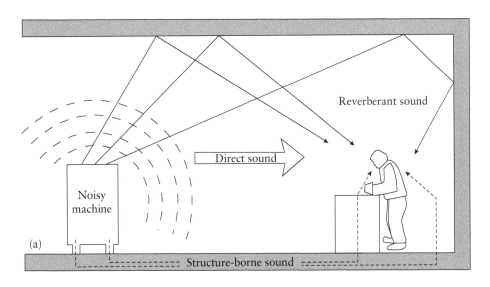

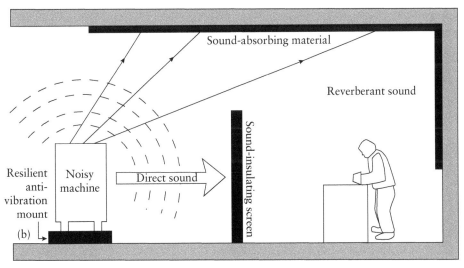

Figure 2.2 (a) Transmission of noise by different paths and (b) methods of noise control. Reproduced with permission from R.J. Peters in *Noise Control (A Pira Environmental Guide)*, Pira, Leatherhead, UK, 2000. ©2000, Pira

The source–path–receiver model will obviously become more complicated still if there are multiple sources as well as different paths, and the complexity will be increased further if different receivers are also considered. Therefore, correct diagnosis of sources and paths become an important aspect of noise control.

2.5 Noise Control at Source

2.5.1 Machinery Noise – Generation and Control

The source–path receiver model may also be applied to the analysis of noise radiated from machines. The source in this case is the noise-producing mechanism, which is often a dynamic force, producing vibration that is transmitted *via* the structure of the machine to the external surfaces, which radiate the noise. An example is the combustion forces occurring within the cylinder of an internal combustion engine, which are transmitted through the engine structure to noise-radiating surfaces such as crankcase panels and thin sheet metal covers (e.g., the sump and rocker cover). Often, the main structural framework of the machine may be strong and stiff and designed to withstand the working forces generated within the machine. However, most of the noise is generated by much more flexible sheet steel covers attached to the external parts of the frame. A good example would be a punch press.

A full understanding of the noise-generating forces and mechanisms, and the way in which they are transmitted to the noise-radiating surfaces, is essential if the noise is to be reduced at source by modification to the design of the machine.

Apart from devising an entirely different process, the opportunities for reduction are in reducing or changing the vibration-producing forces, changing the transmission paths by which these forces reach the radiating surfaces, or by changes to these surfaces.

The forces are often the same as the working forces that perform the machine function. They include:

- Cutting, pressing and shaping forces;

- Gear and bearing forces;

- Electric and magnetic forces (in motors and generators);

- Combustion forces;

- Out of balance forces (in rotating machines);

- Frictional forces, for example, in brakes;

- Pneumatic and hydraulic forces, for example, in pumps; and

- Stick-slip forces which can result in the 'squeal' of brakes and cutting tools.

Research has been carried out to modify these forces without reducing the working efficiency of the machines. Thus, it is now possible to obtain machines with quieter gears and bearings, or to replace gears by belt drives, better balanced rotating machines and quieter motors. In many cases, the modifications involve better-quality engineering and hence more expense. It is possible to design quieter cutting tools for presses and other material-forming tools.

The rate of change of force is important, as well as its magnitude. Smooth forces that change slowly are quieter than those that vary rapidly. A good example is the sharp rate of increase in combustion pressure which occurs in diesel engines, which gives rise to the characteristic 'diesel knock' which is absent from the noise from petrol engines. There are ways of modifying the combustion forces that reduce noise but they also have effects on gaseous emissions (e.g., hydrocarbons, oxides of nitrogen) and on fuel efficiency. Another example is the cutting of sheet metal using a guillotine which, because the cutting force is applied smoothly and steadily, is a very quiet process, although the impact of the offcut onto the floor is often noisy unless measures are taken to 'cushion' the blow.

Figure 2.3 shows a typical noise spectrum from hammering activity and the effect of using a sound-deadened hammer.

Once the vibration has reached the vibrating surface several measures can reduce the noise:

- To reduce the surface area of the noise-radiating surface;

- To reduce the forces by isolating the panel from the frame of the machine;

- Stiffening or damping the panel; and

- Close shielding or applying acoustic lagging to the panel.

If the surface is relatively stiff, a cast panel, for example, then damping which relies on large vibration amplitudes will be ineffective, and isolating the panel may be the best option. If the surface is a flexible thin sheet steel panel, damping will be an appropriate treatment. A possible alternative would be to stiffen the panel by increasing its thickness or by attaching stiffening ribs and, if necessary, isolating the stiffened panel from the frame. The effect of increasing stiffness may be complex and difficult to predict. Changing the stiffness changes the natural frequencies of the panel and also its radiation efficiency.

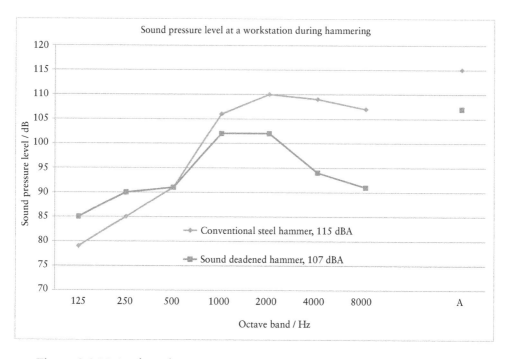

Figure 2.3 Noise from hammering activity at a workstation. Adapted from BS EN ISO 11690-2:1997 Acoustics – Recommended practice for the design of low-noise workplaces containing machinery – Part 2: Noise control measures, Figure 4

In many material-forming processes, the workpiece being formed may also radiate a significant amount of noise. When a blacksmith's hammer strikes a metal bar on an anvil, then the noise generated by the impact will be radiated by the hammer, the workpiece, the anvil and the bench to which the anvil is attached. Similarly, timber planks passing through woodworking machinery will also radiate noise, as will sheets of metals being drilled, punched or riveted, or metal castings being dressed by chisels, grinders or chipping hammers, or machined by lathes or milling machines. Noise radiated from workpieces can be reduced by clamping or supporting the workpiece to provide temporary increased damping and reduce vibration. Hollow castings may be filled with sand to provide damping and reduce the ringing noise during machining; clamping and damping of sheet metal will have a similar effect. Sometimes, enclosures around woodworking machines are fitted with long entrance-and-exit tunnels, lined with sound-absorbing material to enclose the wooden planks while they are machined and to attenuate noise escaping through the work access holes.

The noise levels from most rotating machines increase sharply with increasing speed. Therefore, any changes that reduce machine speed will provide a useful reduction in noise. Large, slow-speed fans, for example, are usually much quieter than smaller, high-speed versions which move similar volumes of air.

The level of noise from pneumatic jets and nozzles used widely for cleaning purposes in industry increases rapidly with air pressure, and any reduction in pressure will lead to significant noise reductions.

Good maintenance will reduce noise caused by the movement of worn or loose parts and correct lubrication will reduce noise from bearings and gears. Keeping the cutting edge of tools sharp will also reduce noise as well as maintain the efficiency of the process.

2.6 Noise Control during Sound Transmission

The concepts of sound insulation, sound absorption (for the control of airborne sound), and vibration isolation and damping (for control of vibration and structure-borne sound) were introduced in **Chapter 1**. Good sound insulation is required for: noise-reducing partitions, screens and barriers; the close shielding or acoustic cladding of noise-radiating surfaces; partial and complete machine enclosures. Sound-absorbing materials are used: to reduce the levels of reverberant sound in spaces as linings to walls floors or ceilings or freely hanging mineral fibre slabs; in absorptive silencers for reducing noise in ducts and in the cavity of double-leaf partitions; as sound-absorbing linings for screens barriers and acoustic enclosures. The use of anti-vibration mounts to prevent the transmission of vibration from machines to adjacent surfaces was explained in principle in **Chapter 1**, and is discussed in more detail below.

2.7 Isolation

The types of resilient materials that may be used for vibration isolation are: various types of resilient pads or mats; rubber or neoprene of similar polymeric materials; metal springs. Vibration-isolating pads are made from cork or felt, mineral fibres or plastic foam pads. Their resilience arises from the fact that they all contain air in pores between fibres and they are used in compression. By contrast, rubber becomes flexible when it changes its shape and is used in shear-vibration isolators (**Figure 2.4**).

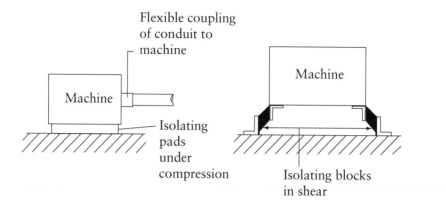

Figure 2.4 Use of isolating pads under compression and rubber blocks used in shear (schematic)

Figures 2.5 and **2.6** show examples of anti-vibration mounts featuring rubber in shear and metal springs, respectively.

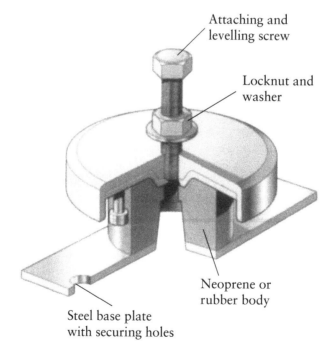

Figure 2.5 Anti-vibration mount using rubber or neoprene in shear. Reproduced with permission from Controlling Noise at Work L108 under *the Control of Noise at Work Regulations 2005*, Guidance on the Regulations, Health and Safety Executive, Bootle, Merseyside, UK, 2005. ©2005, Health and Safety Executive

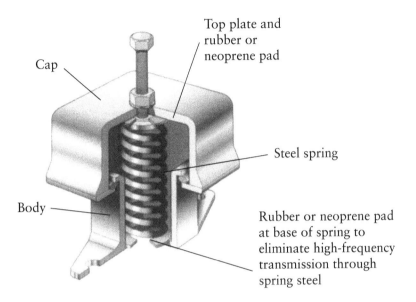

Cap

Top plate and
rubber or
neoprene pad

Steel spring

Body

Rubber or neoprene pad
at base of spring to
eliminate high-frequency
transmission through
spring steel

Figure 2.6 Anti-vibration mount using metal springs. Reproduced with permission from Controlling Noise at Work L108 under *the Control of Noise at Work Regulations 2005*, Guidance on the Regulations, Health and Safety Executive, Bootle, Merseyside, UK, 2005. ©2005, Health and Safety Executive

As well as being used to isolate a vibrating machine from the building structure, resilient material can be used to isolate some part of a machine from another, for example, to prevent the vibration from a generator or a fuel pump from an engine structure. Machine panels and covers can be isolated from machine structures (**Figure 2.7**). **Figure 2.8** illustrates the sort of design detail needed to ensure that the isolation is not bridged, for example, by connecting the fixing bolts that connect the panel to the machine frame.

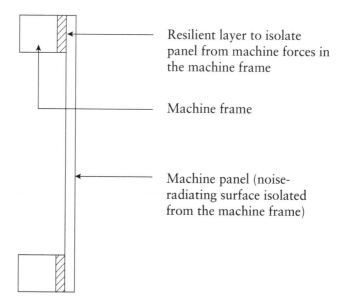

Resilient layer to isolate panel from machine forces in the machine frame

Machine frame

Machine panel (noise-radiating surface isolated from the machine frame)

Figure 2.7 Isolation of noise-radiating panels from a machine frame. Reproduced with permission from R.J. Peters, B.J. Smith and M. Hollins in *Acoustics and Noise Control*, 3ʳᵈ Edition, Prentice Hall, Upper Saddle River, NJ, USA, 2011. ©2011, Prentice Hall

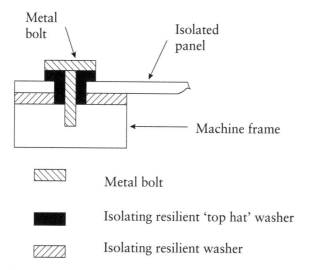

Metal bolt

Isolated panel

Machine frame

Metal bolt

Isolating resilient 'top hat' washer

Isolating resilient washer

Figure 2.8 Use of an isolating washer to prevent the bridging of isolation between a machine frame and the outer panel. Reproduced with permission from R.J. Peters, B.J. Smith and M. Hollins in *Acoustics and Noise Control*, 3ʳᵈ Edition, Prentice Hall, Upper Saddle River, NJ, USA, 2011. ©2011, Prentice Hall

2.8 Damping

'Damping' is a mechanism or process whereby vibrational energy in a material is converted into heat by some form of frictional mechanism. Certain materials, for example, mild steel, have very little internal damping and will ring like a bell if struck. Other materials, such as timber and some forms of plastic, have much more internal damping.

Increasing the amount of damping in a sheet or panel of a machine will be effective for reducing noise and vibration in two circumstances (**Figure 2.9**). First, if the sheet or panel is subjected to a series of impacts and is relatively undamped, each impact will produce a ringing noise and this may be reduced by using a material with more damping, or by applying damping material to the panel. Examples where this application would be suitable would be for panels used as bins or hoppers or chutes in the handling and transport of material.

The second application is if the panel is subject to resonant forced vibration. This could be the case, for example, if a sheet steel gearbox cover or fan casing is resonating because the driving speed produces a frequency of forcing vibration that coincides with the natural frequency of the panel. In this case, increased damping will reduce the amplitude of the resonant vibration and hence the noise.

The amount of damping of a sheet steel panel may be increased by applying a layer of highly damped material, for example, some sort of viscoelastic polymer (**Figure 2.10**). This is called 'unconstrained layer damping'. The vibration of the sheet steel panel is transmitted to the polymer layer and the damping mechanisms within this layer extract sound and vibrational energy. A more effective treatment is to apply a second constraining sheet on top of the damping layer, thus making a 'sandwich': this is called 'constrained layer damping'.

Several proprietary products are available for use as damping materials. They can be painted on as a mastic or applied as a stick-on sheet. Sheets of sandwich materials employing constrained layer damping are also available.

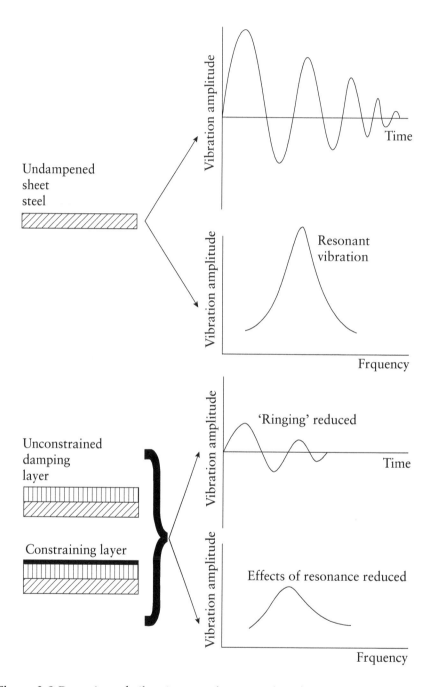

Figure 2.9 Damping of vibrating panels. Reproduced with permission from R.J. Peters in *Noise Control (A Pira Environmental Guide)*, Pira, Leatherhead, UK, 2000. ©2000, Pira

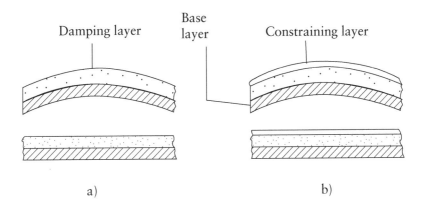

Figure 2.10 (a) Unconstrained layer and (b) constrained layer damping. Reproduced with permission from R.J. Peters, B.J. Smith and M. Hollins in *Acoustics and Noise Control*, 3rd Edition, Prentice Hall, Upper Saddle River, NJ, USA, 2011. ©2011, Prentice Hall

2.9 Close Shielding or Acoustic Lagging

Sometimes, rather than attempting a complete enclosure around a machine, it is better to shield only those surfaces known to radiate most of the noise, for example, thin sheet steel panels. An arrangement that is suitable is shown in **Figure 2.11**.

The shield should consist of two layers: a resilient layer next to the noise-radiating surface and a heavy, massive layer. The resilient layer may be of mineral fibre or foam. The thickness of the resilient layer should be 25 to 50 mm depending on the noise reduction required and the frequency of the sound to be shielded. The mass layer could be of sheet steel, lead sheets, plasterboard, mortar, or of timber (e.g., fibreboard or chipboard). The noise-reduction process is complex, but the major mechanism is probably the effect of isolation. The resilient layer and the mass layer form a mass–spring system that isolates the mass layer from the noise-radiating surface effectively. However, there may also be an effect of sound absorption, damping, and sound insulation. The treatment is most effective at high frequencies and it is difficult to obtain high-noise reductions at low frequencies.

As well as being used for reducing noise from panels, this treatment is also suitable for the acoustic lagging of pipes and ducts (**Figure 2.12**). Attenuations of less than 6 dBA are estimated from this type of treatment but, depending on the frequency content of the noise, higher attenuation will usually be achieved for higher frequencies. Structural contact between the vibrating surface and external mass layer must be avoided.

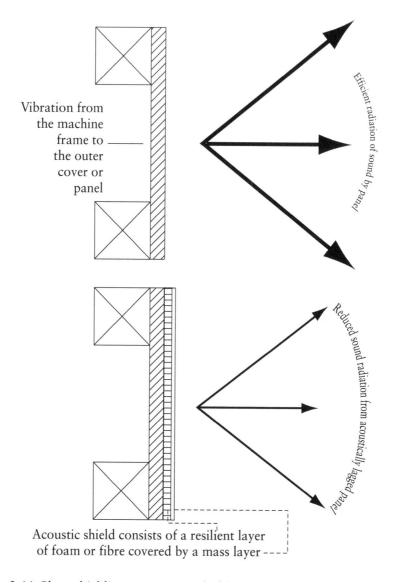

Vibration from the machine frame to the outer cover or panel

Efficient radiation of sound by panel

Reduced sound radiation from acoustically lagged panel

Acoustic shield consists of a resilient layer of foam or fibre covered by a mass layer

Figure 2.11 Close shielding or acoustic cladding (or lagging) of noise-radiating panels. Reproduced with permission from R.J. Peters in *Noise Control (A Pira Environmental Guide)*, Pira, Leatherhead, UK, 2000. ©2000, Pira

If more attenuation is needed, a second layer of acoustic lagging may be used, as shown in **Figure 2.13**, for which noise reductions between 10 and 25 dBA might be achieved, again depending on the frequency content of the noise being radiated by the pipework.

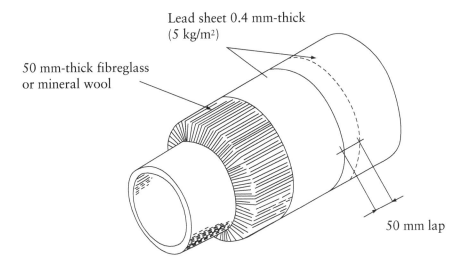

Lead sheet 0.4 mm-thick
(5 kg/m²)

50 mm-thick fibreglass
or mineral wool

50 mm lap

Figure 2.12 Reduction of noise from a radiating surface by external lagging (single stage). Reproduced with permission from *Noise in the Plastics Processing Industry: A Practical Guide to Reducing Noise from Existing Plant and Machinery*, Smithers Rapra, Shawbury, UK, 1985. ©1985, Smithers Rapra

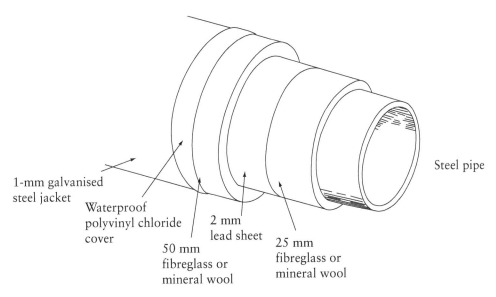

Steel pipe

1-mm galvanised
steel jacket

Waterproof
polyvinyl chloride
cover

2 mm
lead sheet

50 mm
fibreglass or
mineral wool

25 mm
fibreglass or
mineral wool

Figure 2.13 Acoustic lagging or cladding (two stage).
Reproduced with permission from *Noise in the Plastics Processing Industry: A Practical Guide to Reducing Noise from Existing Plant and Machinery*, Smithers Rapra, Shawbury, UK, 1985. ©1985, Smithers Rapra

2.10 Use of Screens and Barriers

A screen or barrier placed between a sound source and receiver will reduce the direct sound transmission between the source and receiver, by an amount which depends on the sound-insulating quality [i.e., the sound reduction index (R)] of the barrier. In an outdoor situation where there may be no nearby reflecting surfaces to direct sound around the barrier towards the receiver, attenuations of less than 10 dB may be achieved. The use of barriers outdoors is discussed in **Chapter 7**, including how attenuation may be predicted.

If noise barriers are used indoors, sound reflections may reduce the effectiveness of the barrier (**Figure 2.2**) and reductions will be lower, may be limited to 5 or 6 dBA depending on the frequency of the sound.

To maximise the effectiveness of an indoor noise barrier, it should be:

* As close to the source as possible;

* Shaped to wrap around the source as far as possible;

* Lined with sound-absorbing material on the source side; and

* Nearby reflected surfaces should also be lined with sound-absorbing materials.

Some of these features are illustrated in **Figure 2.14**.

An acoustic enclosure is a complete box (i.e., four sides and a roof) built around a machine to contain or confine noise from the machine within the enclosure, and so to reduce the noise transmitted to the surrounding area. A well-designed acoustic enclosure employs all three principles of insulation, absorption and isolation.

The principles of design are threefold. First, the walls of the enclosure should have a sufficient sound reduction index to achieve the desired noise reduction. Second, it is advisable to line the inside of the enclosure with sound-absorbing material. This will reduce the noise level inside the enclosure and, thus, supplement the effect of the sound insulation by further reducing the level of sound transmitted through the enclosure walls to the outside. If the enclosure is to be large enough for people to work inside it, then this sound-absorbing lining is essential because without it the level inside the enclosure will be higher than before the enclosure was built because of a build-up of reverberant sound. The third design principle is that the machine inside the enclosure should be isolated from the floor to reduce structure-borne sound transmission. Any service pipes or ducts fitted to the machine must be connected flexibly at the machine and pass through the walls of the enclosure with a flexible seal so that structure-borne sound is not transmitted *via* the enclosure walls.

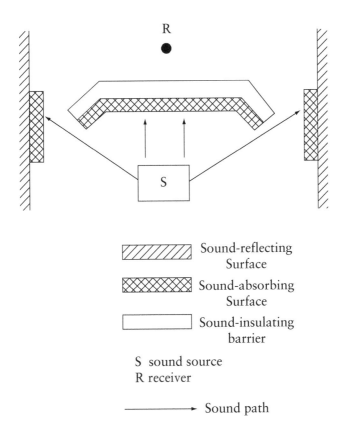

Sound-reflecting
Surface

Sound-absorbing
Surface

Sound-insulating
barrier

S sound source
R receiver

Sound path

Figure 2.14 Use of an indoor barrier between the sound source and receiver

In most cases, holes and gaps in the enclosure will be needed to feed services and materials into and out of the machine. The effectiveness of the enclosure will depend upon the extent to which the effect of these apertures can be reduced so that the potential noise reduction obtainable from the material of the enclosure is achieved.

The use of acoustic enclosures as noise-control devices has several advantages and disadvantages. The main advantage is that the concept of an acoustic enclosure is extremely simple and can be adapted to almost any machine or noise source. Provided that a material of sufficient density and thickness is used, acoustic enclosures may be designed to achieve almost any reasonable amount of noise reduction: up to 30–40 dBA. At one extreme, enclosures may be made of brick and designed to reduce high levels of low-frequency noise from large diesel generators set close to noise-sensitive areas. At the other extreme, they may be made of thin plastic sheet and designed to reduce high-frequency noise from an office printer. **Figure 2.15** illustrates the main features of a successful acoustic enclosure:

- A sound-insulating enclosure wall (typically 16 or 18-gauge steel).

- A sound-absorbing enclosure lining (typically 2–4 inches of mineral fibre, faced on the inside with 22-gauge perforated sheet steel) to reduce the build-up of reverberant sound inside the enclosure.

- Anti-vibration mounts beneath the machine feet to isolate them from the floor and prevent structure-borne sound transmission.

- Apertures in the enclosure for pipes and ducts carrying materials and services to the machine should be well sealed using resilient materials to prevent bridging of the isolation.

- Inspection hatches, doors, and windows must be constructed to the same standard as that for the rest of the enclosure, and must be well sealed.

- Ventilation should be *via* ducts lined with sound-absorbing material to minimise the scale of noise *via* ventilation apertures.

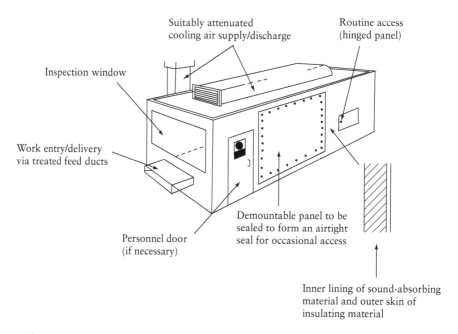

Figure 2.15 An acoustic enclosure showing typical panel construction. Reproduced with permission from Controlling Noise at Work L108 under *The Control of Noise at Work Regulations 2005*, Guidance on the Regulations, Health and Safety Executive, Bootle, Merseyside, UK, 2005. ©2005, Health and Safety Executive

One of the disadvantages of acoustic enclosures is that they can interfere with the normal operation of machines, for example, if operators have to feed or remove material, or change settings on the machine. Another disadvantage is that many machines will become unduly heated when enclosed unless provision is made for air to circulate inside the enclosure and, if forced ventilation is necessary, this will require a fan, which will produce noise and require acoustic treatment. Acoustic enclosures may be relatively expensive and take up valuable floor space.

Partial acoustic enclosures (i.e., with one or more sides removed) will be less effective than full enclosures but can give less than 10 dBA noise reduction if lined with sound-absorbing material.

Figure 2.16 illustrates the relative importance of sound-absorbing linings in the peformance of the enclosure. The enclosure without acoustic lining and with open ventilation openings but with the machine isolated from the floor produces a reduction of 7 dB (comparison between parts **Figure 2.16a** and **b**). The anti-vibration mounts produced a reduction of 2–3 dB reduction at low frequencies (31.5, 63 and 125 Hz octave bands) but had no effect on the dBA level.

Fitting ventilation ducts with sound-absorbing lining produces an additional reduction of 11 dB (i.e., 18 dBA overall [comparison of **Figure 2.16b** and **c**]).

Lining the sound-absorbing material with sound-absorbing material produces an additional 6 dBA reduction (i.e., 24 dBA overall [comparison of **Figure 2.16c** and **d**]).

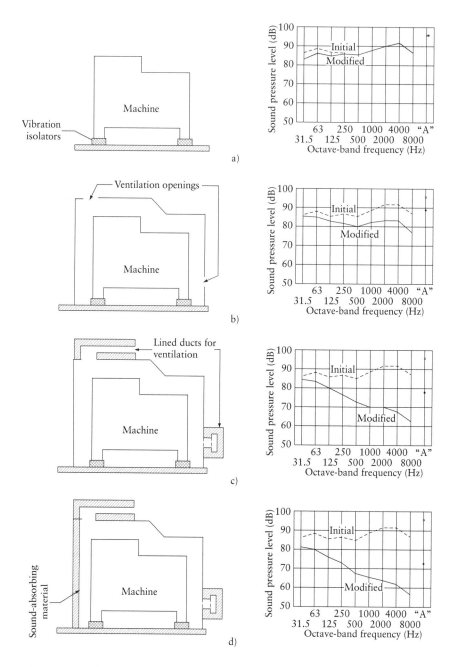

Figure 2.16 Typical noise reductions achieved for different machine and enclosure configurations. Reproduced with permission from BS EN ISO 11690-2:1997 Acoustics – Recommended practice for the design of low-noise workplaces containing machinery – Part 2: Noise control measures. ©1997, British Standards Ltd

2.11 Specification and Selection of Damping, Isolating and Lagging Materials

As discussed in **Chapter 1**, there are standard test methods for determining the sound reduction index of sound-insulating building elements, and for the absorption coefficient (α) of sound-absorbing materials.

There are standard design procedures for selecting anti-vibration mounts to place under machines to prevent the transmission of 'feelable' vibration to the floor and building structure. These involve the load to be borne by the isolators and its stiffness, and usually specified in terms of a static deflection of the isolators due to the weight of the machine or to the natural frequency of the isolators and machine acting together as a 'mass spring' vibrating system.

There is much less guidance available for selecting pads, sheets and wrapping layers to be used for isolating material to prevent the transmission of structure-borne noise as well as for damping and acoustic lagging or cladding purposes.

The standard, ISO 15665:2003 *Acoustics – Acoustic insulation for pipes, valves and flanges*, describes a method for specifying, designing and testing the acoustic insulation of pipework from 300 mm up to 1 m in diameter, but it is not applicable to smaller-diameter pipework or to rectangular ducting and vessels or machinery.

There is a large variety of acoustic material products available commercially, some of which are clearly designed for damping or isolation applications and others as acoustic lagging. However, many of them offer a combination of properties that combine the effects of sound insulation, absorption, isolation and damping. Manufacturers' data sheets usually offer test-performance data, perhaps showing the noise reductions achieved for a particular application. These should be scrutinised carefully to see to what extent the test situation resembles the proposed application.

Ideally, any data on noise-reduction performance should be available in octave bands. The user should beware of claims for overall reduction in dBA because, although they may be given in good faith, they can only at best be for 'typical' cases. The overall reduction in dBA obtained in each case will depend on the frequency spectrum (in octave bands) which is to be reduced. If the frequency spectrum of the application is different from the typical case, then a different dBA reduction will result – usually performance is better at high than at low frequencies.

2.12 Fans and Blowers

Fans of one sort or another are responsible for about 30% of complaints about noise nuisance from industry and commerce. They are used for moving air in ventilation systems, removing fumes and waste gases, and for transporting particulate material in the form of process material (powders, granules) or waste particulate material. The axial and centrifugal types used widely for moving air and waste gases produce a broadband noise but tonal components can be produced by interaction with obstacles such as guide vanes placed too close to fan blades. For other types, such as propeller fans and high-pressure industrial blowers for transporting particulate material, the tonal component may be much more prominent.

The control of noise from fans starts with the selection of the fan. The main criterion for selection is the volume flow rate of air that must be moved and the pressure drop against which it must work. Within these parameters, a fan with the lowest noise output when working at its optimum efficiency should be selected. This may need to take into account the shape of the noise spectrum because it may be better to accept a fan that has the lowest sound power level at lower frequencies (which are difficult to reduce) even though it may produce higher levels at more easily controlled high frequencies.

The size of the fan is also an important factor because, in general, larger fans operating at lower speeds move the same amount of air with less noise than smaller, higher-speed fans. The noise output increases rapidly with speed so that if variable-speed fans can be used, this enables noise to be reduced by reducing the speed when the load requirement is low. This is important because fan noise increases rapidly with increasing fan speed, so that, for example, a reduction of 20% in fan speed can cause a noise reduction of 5 dB, and a 50% reduction in fan speed can lead to a 15 dB reduction in noise level from the fan.

Once the fan has been selected it should be installed so as to make the minimum noise possible. This requires a smooth flow of air into the fan because input turbulence can increase the noise level by several decibels. There should be no obstructions to flow, no bends or changes of cross-section within at least 2–3 duct diameters up- or down-stream of the fan (**Figure 2.17**).

Noise from the inlet and supply side of the fan may be reduced by using in-line silencers. Also, airborne noise radiating from the fan casing itself may be reduced by damping, close shielding, or by an acoustic enclosure around the fan. The transmission of structure-borne sound should be reduced by fitting anti-vibration mounts between the fan and the building structure (usually the floor) and using flexible connectors between the fan and ductwork. The sound power produced by the fan may be augmented by noise generated by the flow of air through the duct system, and this should be minimised by designing the system to keep the airflow as smooth and slow as possible. If the airflow carries particulate material, then there is additional noise generated by the impact of the particles with the duct walls. Noise generated within and emanating from the duct system may be reduced using a secondary silencer, and noise breaking out from the duct through the duct walls may be reduced by application of acoustic lagging of the duct.

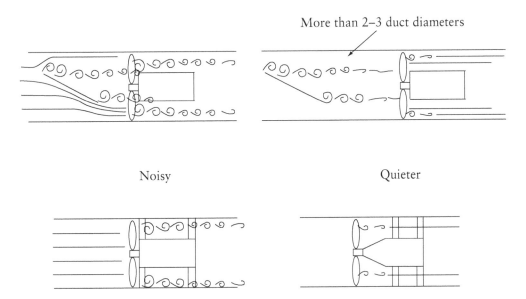

Figure 2.17 Good (quiet) and bad (noisy) fan-installation arrangements. Reproduced with permission from Industrial Noise & Vibration Centre (INVC). ©INVC

In summary, noise problems from fans may be minimised by:

- Careful location of the fan, inlet and outlets to avoid noise-sensitive areas;

- Planning and prediction of noise levels before installation, if possible, so that noise control can be incorporated at the design stage rather than retrospectively;

- Careful selection of the most appropriate type and size of fan, operating at its lowest possible speed;

- Reduction of fan speed, if possible, at periods of reduced load and at sensitive times (e.g., at night);

- Correct installation to ensure that the fan is operating at maximum efficiency;

- Design of a duct system to ensure smooth airflow into the fan and smooth airflow at the lowest possible speed through the ductwork to minimise additional noise generation; and

- Use of silencers, fan enclosure and acoustic lagging of ductwork, if necessary, to provide additional attenuation.

2.13 Acoustic Attenuators or Silencers

Whenever sound travels with a ducted air or gas supply then it is possible to reduce the noise transmitted *via* the duct by using a lining of sound-absorbing material. The principle is that the sound will travel to and fro across the duct and that each time it strikes the wall of the duct some of the acoustic energy will be absorbed by the lining. The principle of a lined duct may be extended to an acoustic attenuator or silencer in which, in addition to lining the sides of the duct sections of sound-absorbing material (often called 'splitters' because they separate the airway in the duct into sections), are inserted into the duct to increase the surface area of the sound-absorbing material and thus the amount of noise reduction achieved. **Figure 2.18** illustrates the structure and **Figure 2.19** shows the typical performance of an acoustic attenuator. The sound-absorbing slabs of material are usually protected against abrasion by thin perforated metal sheets and sometimes also by thin plastic films to prevent fibre shedding, protect against moisture, or for hygiene reasons.

The amount of attenuation provided by the silencer will depend upon the:

- Absorption coefficient of the sound absorbing material;

- Thickness of the sound-absorbing linings;

- Number and thickness of the splitters;

- Width of the airway passages between the sound-absorbing slabs (splitters); and

- Length of the silencer.

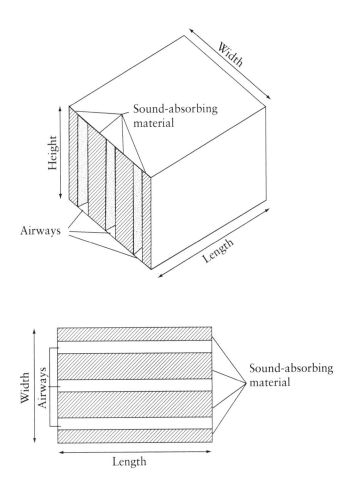

Figure 2.18 Main features of an absorptive splitter silencer.
Reproduced with permission from R.J. Peters, B.J. Smith and M. Hollins in
Acoustics and Noise Control, 3rd Edition, Prentice Hall, Upper Saddle River, NJ,
USA, 2011. ©2011, Prentice Hall

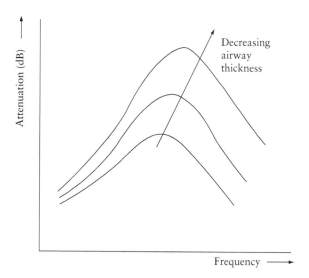

Figure 2.19 Acoustic attenuation performance of an absorptive splitter silencer. Reproduced with permission from R.J. Peters, B.J. Smith and M. Hollins in *Acoustics and Noise Control*, 3rd Edition, Prentice Hall, Upper Saddle River, NJ, USA, 2011. ©2011, Prentice Hall

The absorption coefficient of the splitters, and thus the attenuation produced by the silencer, increases with increasing frequency and, at low frequencies in particular, depends upon the thickness of the splitters. At high frequencies, the attenuation falls off because the sound becomes more directional and tends to travel down the airways and be less affected by the sound-absorbing material linings. The attenuation is, therefore, improved by reducing the width of the airways, but this will increase the flow resistance of the silencer, thus requiring more energy (and more noise) from the fan to maintain the flow. Restricted airway thickness can also lead to the creation of noise generated by airflow through the silencer. The design of the silencer is, therefore, a compromise between attenuation, pressure drop, flow-generated noise, size (length and cross-sectional area) and material cost.

The type of silencer described above is called an 'absorptive' or 'dissipative' silencer because it relies on the properties of sound-absorbing materials. It produces attenuation over a wide range of frequencies, but is least effective at low frequencies (when very large and expensive units are required to obtain significant noise reductions).

'Tuned' or 'resonant' sound absorbers form the basis of 'reactive' silencers which may be used to obtain large attenuation over a narrow range of frequencies. They utilise the principles of resonance and of interference between the waves reflected at different points in the system. A simple example is an expansion chamber, for which the attenuation is maximum when the length of the chamber corresponds to one-quarter of the wavelength of the sound. The useful range of frequencies can be extended by combining chambers of different length, and also by introducing pipes between the different sections so that additional resonances and anti-resonances are introduced related to reflections in the pipes. **Figure 2.20a–c** shows the typical construction and sound-attenuation performance of reactive silencers based on the principle of an expansion chamber.

The Helmholtz resonator is perhaps the best-known example of a reactive attenuator. It consists of a volume or cavity connected to a short open neck. The device acts as a simple mass–spring system with a sharply defined natural frequency. Blowing across the top of a partly full milk bottle illustrates the principle. The sound wave causes the air in the neck (the mass) to vibrate and cause pressure fluctuations in the cavity (the spring). At the natural frequency, the cavity absorbs large amounts of energy from the sound wave. A similar effect is produced when a series of holes in a pipe allow the air or gas flowing through the pipe to communicate with a chamber surrounding the pipe, or by a closed side-branch pipe or duct that produces a maximum attenuation when the length of the side branch is one-quarter of a wavelength. These various devices and their typical performance are illustrated in **Figure 2.20d**. In many cases, they adopt a combination of absorptive and reactive principles. Reactive attenuators are used for low-frequency sounds where solely absorptive types are relatively ineffective, and they are particularly useful for reducing the level of low-frequency pure tones caused by machinery operating at low and fixed speeds because the dimensions of the device can be designed to produce maximum attenuation at the frequency of a particular tone. They are used widely for silencing vehicle exhausts, compressors, pumps, and fans that produce pure tones.

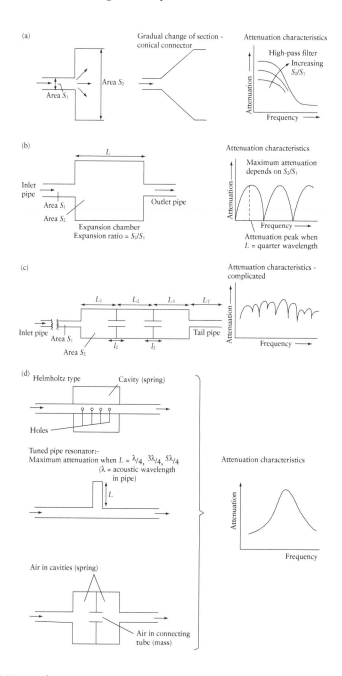

Figure 2.20 Typical construction and sound-attenuation performance of reactive silencers based on the a–c) expansion chamber principle and d) Helmholtz resonator principle. Reproduced with permission from R.J. Peters, B.J. Smith and M. Hollins in *Acoustics and Noise Control*, 3rd Edition, Prentice Hall, Upper Saddle River, NJ, USA, 2011. ©2011, Prentice Hall

2.14 Jets and Exhausts

A jet is a flow of fast-moving gas or air moving from the end of a tube or pipe into free air. The structure of a jet is shown in **Figure 2.21**. The noise-producing mechanism is the turbulence created by the viscous forces which operate between adjacent layers of air in the boundary layer between the jet and surrounding air, where there is a very high velocity gradient. The sound power level produced depends on the cross-sectional area of the jet and the efflux velocity. The level increases sharply with increasing velocity so that any velocity reduction which can be achieved will produce significant noise reductions: a reduction of more than 20 dB if the velocity is halved. The noise has a broadband spectrum with a peak at a frequency that is directly proportional to the jet velocity, v (in m/s), and is inversely proportional to the jet diameter, d (in mm). Hence, small-diameter high-velocity jets will produce predominantly high-frequency noise and, conversely, large-diameter and low-velocity jets will be predominantly low frequency noise.

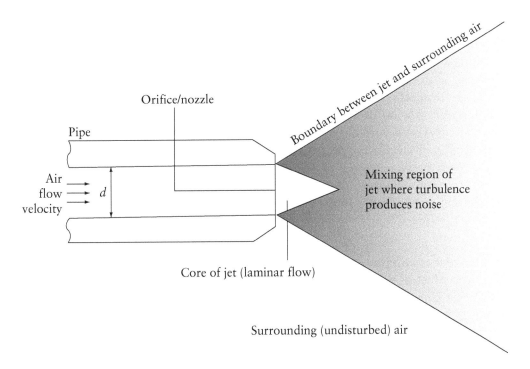

Figure 2.21 Jet structure. Reproduced with permission from R.J. Peters in *Noise Control (A Pira Environmental Guide)*, Pira, Leatherhead, UK, 2000. ©2000, Pira

The noise radiated by a jet is directional and so some benefit can be obtained by directing outdoor jets away from sensitive areas, for example, vertically upwards.

Jet noise is produced in exhausts, vents, valves and other pressure-release devices where air, gas or steam at high pressures and velocities are released into the atmosphere. Jet noise is also produced by the exhausts of pneumatically operated actuators and hand tools, and by 'working' air-jets used to eject workpieces and for cleaning and cooling purposes. Compressed air hoses are among the most universal high-level noise sources in industry.

2.14.1 Pneumatic Jet Silencers

The principles of diffusion and absorption are used to reduce noise from exhaust jets. **Figure 2.22** shows examples of an exhaust silencer suitable for process exhausts in which the initial high-flow velocity is forced through a perforated diffuser. This has the effect of reducing flow velocity, promoting frictional effects, and also converting one single jet which, for a large-diameter process exhaust system may be predominantly low-frequency noise, to many smaller high-frequency jets which are more easily attenuated by the sound-absorbing material lining the silencer. Exhaust silencers working on the diffuser principle, which can be screwed onto pneumatic exhausts easily, are relatively cheap and readily available, and can produce reductions of less than 25 dBA. The diffusing medium may be sintered plastic, compacted wire wool or layers of woven wire cloth. In addition to the noise reduction to be achieved, the selection of the most appropriate type will need to take into account the effects of any back pressure caused by the silencer, as well as arrangements for cleaning, maintenance or replacement to avoid blockage of the diffuser by debris in the airflow. An alternative to fitting a diffusing silencer is to lead the exhaust away through a length of flexible tubing to exhaust in a less noise-sensitive area.

If the air-jet is required to do some useful work, various 'low noise' nozzles are available. Most of these work on the 'air entrainment' or 'induced flow' principle, in which a flow of air of intermediate velocity is introduced into the boundary layer between the fast-moving jet stream and the stationary surrounding air to reduce the velocity gradient, and so to reduce turbulence and noise generation (**Figure 2.23**).

Fabricated and assembled metal body with layers coarse-woven wire cloth as silencing materials

Machined metal body and thread with compacted wire wool as silencing material

Injection-moulded plastic body and thread with sintered plastic or metal membrane

Figure 2.22 Typical silencers for reduction of pneumatic exhaust noise. Reproduced with permission from Controlling Noise at Work L108 under *the Control of Noise at Work Regulations 2005*, Guidance on the Regulations, Health and Safety Executive, Bootle, Merseyside, UK, 2005. ©2005, Health and Safety Executive

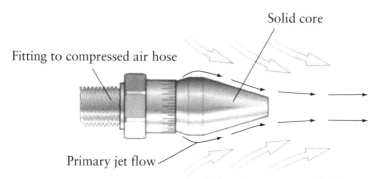

Fitting to compressed air hose

Solid core

Primary jet flow

Induced secondary air flow

Figure 2.23 The principle of an entrained airflow jet silencer. Reproduced with permission from Controlling Noise at Work L108 under *the Control of Noise at Work Regulations 2005*, Guidance on the Regulations, Health and Safety Executive, Bootle, Merseyside, UK, 2005. ©2005, Health and Safety Executive

More detailed information on reducing noise from a pneumatic air-jet is given in the Health and Safety Executive (HSE) document *Noise from Pneumatic Systems: Guidance Note PM 56*.

2.15 Punch Presses

Punch presses are used throughout the engineering industry and can produce high levels of noise and vibration. In addition to the impact noise produced by the action of the die on the workpiece, there are many other sources of noise:

- Compressed air-jet noise;

- Transmission of vibration to external noise-radiating surfaces of the press, and to other surfaces;

- Noise from ancillary equipment, fans, motors, and gears;

- Noise from workpiece feed mechanisms; and

- Impact noise from the collection of workpieces and offcuts.

Some possible noise-control treatments that might be used include: total enclosure; partial enclosure; close shielding/acoustic lagging of existing panels, or replacement by 'acoustic' panels with good sound-insulating properties; exhaust silencers; vibration isolation of the press from the building and of parts of the press from the press frame; damping of panels and of collection bins; modification of the punch tools to reduce impact noise.

2.16 Mechanical Handling and Transport of Materials

In most industries, materials must be moved about and collected using conveyor belts, or airflow in ducts must be used to transport material in particle and pellet form. These processes generate impact noise from materials dropping into chutes and hoppers, and from the pellets striking the walls of ducts, cyclones and collectors. Suitable methods of noise control are to:

- Line the surfaces of conveyors, hoppers and chutes with resilient material to reduce impact noise;

- Use damped material, or apply damping treatments instead of undamped sheet metals for trays, bins, conveyors and hoppers, to reduce ringing noise;

- Reduce impact velocity by reducing the height from which objects drop into hoppers. This can sometimes be achieved by breaking the fall into several smaller stages, with cushioning at each stage; and

- Cover ducts, cyclones and other surfaces conveying pellets with acoustic lagging.

2.17 Active Noise Control

In active noise control (ANC) systems, the principle of destructive interference is applied to cancel the noise using an 'anti-noise' signal from a loudspeaker. The main features of an ANC system installed in a duct are shown in **Figure 2.24**. A microphone picks up the noise signal from the fan, and the signal processor produces a signal of equal amplitude, but of opposite phase, which is fed into a loudspeaker downstream of the microphone. Although the idea is simple in principle it is difficult to carry out in practice. To achieve a reduction of 20 dB, for example, the intensity of the noise must be reduced by a factor of 100, which gives an indication of how exact the cancellation must be. The situation is complicated by the need for a processor to take into account the phase shift introduced by the time it takes for the sound to travel between the microphone and loudspeaker. This depends on the speed of sound, which varies with temperature and with airflow speed. Other problems are that the loudspeaker will also feedback sound to the microphone, as well as cancelling the fan noise, and that the noise from the fan may change with time. These difficulties are overcome by using a second microphone positioned downstream of the loudspeaker to detect any imperfections in the cancellation process and to send an 'error signal' back to the processor, which adapts the cancelling signal to the loudspeaker.

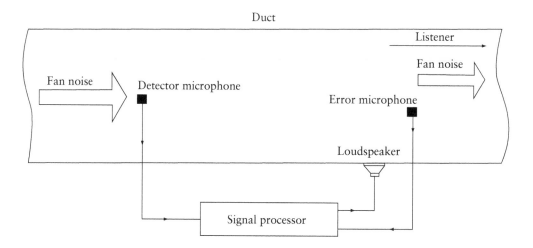

Figure 2.24 ANC applied to fan noise in a duct. Reproduced with permission from R.J. Peters in *Noise Control (A Pira Environmental Guide)*, Pira, Leatherhead, UK, 2000. ©2000, Pira

ANC has been applied successfully to a wide variety of noise sources, including industrial fans, ventilating systems in buildings, generators, compressors and exhaust stack noise. It is best suited to reducing noise levels in small, enclosed spaces, such as the interior of sports cars and aircraft cabins, in ducts and pipes, and in industrial noise havens. Communication headsets and ear muffs fitted with ANC are available. The method is also best suited to the cancellation of low frequencies, for which the wavelengths are long, and so the sound level and phase change gradually with variation in position.

There are some concerns about the ruggedness of ANC devices such as microphones and loudspeakers when exposed to hostile industrial environments for long periods. However, continuing development is likely to lead to improvements in software and hardware, and to new applications.

Alternative 'passive' control methods, such as absorptive silencers, are least effective at low frequencies. Hence, hybrid noise-control systems which use active methods at low frequencies and passive methods at high frequencies are sometimes used.

Figure 2.25 shows ANC loudspeakers built into a passive duct attenuator (splitter silencer). For this particular installation, the ANC produced the following additional attenuation in the low-frequency 1/3 octave bands: 6 dB at 50 Hz; 9 dB at 63 Hz; 13 dB at 80 Hz; 18 dB at 100 Hz; 14 dB at 125 Hz; and 9 dB at 160 Hz.

Figure 2.25 Active noise control loudspeakers in a hybrid active/passive duct attenuator. Reproduced with permission from K. Marriott, ICTC Ltd. ©ICTC Ltd

2.18 Specification of Noise Emission Levels from Machinery, Plant and Equipment

An important part of any hearing-conservation policy for a company should be a requirement to take noise into consideration when selecting and purchasing new and replacement equipment.

It is, therefore, important to have clear, unambiguous and standard methods for specifying the noise output (i.e., noise-emission levels) from the plant and machinery. This may be done by specifying: (i) sound pressure level(s) (L_p) measured at a specified distance from the machine; and (ii) sound power level of the machine.

Machinery noise regulations, issued in response to European Union directives, set out requirements for specifying and declaring noise-emission values for various types of machinery and equipment.

Under these regulations, a series of noise test codes are being developed for each type of machine to indicate how the machine is to be tested, its operating and mounting conditions during the test, the measurement standard to be used, and how the measurement results are to be declared and verified. The use of sound power levels and emission sound pressure levels is recommended in these codes and standards, which are designed to help the manufacturers, suppliers, purchasers and users of machinery.

The machinery noise regulations are described in more detail in **Chapter 8**, which also includes discussion of the relative advantages of using sound power level or sound pressure level at a specified distance, as well as details of the noise test codes for plastics and rubber granulators and shredders.

The importance of drawing up noise specifications for the purchase of new machinery has been emphasised in the HSE booklet *Protection of Hearing in the Paper and Board Industry*. The advice is applicable to all industries. The booklet also emphasises the need for good liaison between the purchaser and supplier of the machinery. The noise specification should include:

- Noise limits for the installation;

- Responsibility of the purchaser and supplier, particularly with regard to the accuracy of information supplied by each; and

- Methods and materials to be used for noise control, including any possible constraints on these because of working practice, environmental, safety or other considerations.

The discussions between purchaser and supplier should include consideration of installation and operating conditions and work patterns for the new plant, noise from any associated new auxiliary plant or existing plant, as well as servicing maintenance arrangements and programmes.

2.19 Diagnosis of Noise Sources, Paths and Mechanisms

Diagnosis is an important first step in any noise-reduction programme. It attempts to answer the following questions:

- Which are the most important noise sources?

- What are the noise-producing mechanisms?

- What are the most important paths of sound transmission?

- Which are the main noise-radiating surfaces?

Good diagnosis enables noise control to be conducted in a systematic manner and the appropriate noise control treatments to be specified and prioritised. Incorrect diagnosis can result in the application of the wrong noise-control measures, leading to a waste of time, effort and money, and bringing the noise-control programme into disrepute.

If there are several different noisy machines, then the best way to find out which ones are the noisiest is to switch them all off and then to measure the noise level with each one being switched on in turn. The alternative method, of measuring the noise while switching machine off at a time, can often be ineffective, as shown by the following example, repeated here from **Chapter 1**.

Ten identical machines each produce a level of 80 dBA at the receiver. Using the rules of combining decibels (**Chapter 1**) it is possible to calculate how the noise level at the receiver increases as each one is turned on: 80, 83, 85, 86, 87, 88, 88.5, 89, 89.5 and 90 dB. This tells us that switching just one source off, leaving the other nine still running, will reduce the total noise by 0.5 dB. Even switching four of the sources off will achieve a reduction of only 2 dB. In this case, the only way to find the contribution from each source will be the first method: switching off all machines and then switching on individual sources singly and in turn. This example also illustrates that the only effective strategy for noise control in this case will be to attempt to reduce the noise levels of all ten sources equally. That is, it would be more effective to reduce the level of all ten sources by 2 dB than to concentrate attention on only two sources and reduce their level by 10 dB each.

Now, let us imagine the situation has changed so that one of the ten sources produces a noise level which is 10 dB above the other nine, which remain equal at 80 dB. The first nine sources switched on together will produce a combined level of almost 90 dB, which will increase to 93 dB when the tenth is also switched on. In this case, the strategy of switching machines off in turn would be successful in diagnosing the one dominant source, but even switching this one off would produce a reduction of only 3 dB, and switching off the others would not produce any detectable difference. This situation would also require an entirely different strategy for noise control. The first step in this case must be to reduce the noise from the one dominant source, by 10 dB, before paying any attention to the other nine. Once the dominant noise source has been treated, the diagnostic exercise must be repeated to establish the new order of priorities for noise control.

Two simple extreme cases have been described to illustrate different strategies that may be required in different situations. In many cases involving multiple noise sources, the choices will not be so clear, but it will still be necessary to draw up a list showing the contributions from each source to the total noise level to prioritise noise-control treatments. The process will need to be ongoing and to be reviewed as the noise-control programme proceeds. In the last example, the priority will be to treat only the dominant source, but once this has been made quieter by 10 dB then, to achieve further noise reductions, the strategy must be switched to treating all ten sources equally.

Sometimes it may be possible to turn off equipment for diagnostic purposes, but only for a very short period of time, perhaps a few seconds. In such cases, it is extremely useful to have made audio recordings of the noise for future replay and analysis. If switching off machines is not possible, then some idea of the relative importance of noise from different machines may be obtained by measuring sound pressure levels close to each machine, where the noise reading will be largely unaffected by that from other nearby machines.

Some machines contain many different sources of noise, for example, a motor vehicle. Noise is produced by the engine, exhaust, air inlet, gearbox and transmission, fan, alternator, tyres and aerodynamically as a result of airflow over the vehicle body. The noise from the engine can in turn be analysed into combustion noise and mechanical noise from valves, pistons, bearings and timing equipment. In addition, vibration from smaller machines, such as fuel pumps and alternators, may be radiated from the engine surfaces to which they are attached, as well as directly.

Although there are specialist acoustic methods for diagnosing sources, the most effective method is for an engineer with a detailed knowledge of how the machine works to conduct a thorough investigation. By walking around the machine, looking and listening carefully, and observing the effect on the noise as the machine's operating conditions (e.g., speed and load) are changed and with various parts of the machine not in operation, one can obtain a good idea of all the important sources and mechanisms. The experience of the operator who uses the machine every day may also be useful. This is rather like detecting the source of a new noise in a car – some indication of the cause may be obtained by listening to the effect of driving at different speeds, in different gears (including neutral) and engine revs, and of accelerating, braking and cornering.

2.19.1 Identifying Sound-Radiating Surfaces

Sometimes it is necessary to find out which surfaces of the machine are radiating the most noise. One effective approach is to cover all possible surfaces with carefully designed and close-fitting acoustic shields or lagging. Then, rather like the method for individual machines, the lagging is removed from each surface in turn (i.e., with all other surfaces covered) and the noise levels are measured to detect which surface makes the biggest contribution to the total. This is the classic method used by the late Professor Priede of Southampton University to detect and rank the important noise-radiating areas of a diesel engine, and has been used by the author to identify a noisy motor as an important contributor to the noise from a large printing plant.

If it is not possible to take valid measurements of machine noise because of the disturbing effect of noise from nearby machines, then it may be possible to measure the levels of vibration on the machine surfaces and estimate the noise level from these. Specialist knowledge is needed, however, to interpret the results. Another specialist method is the use of a sound-intensity meter which, unlike an ordinary sound-level meter, can detect the direction and flow of acoustic energy, and may be used to diagnose which are important sources and radiating areas in a machine, or to measure the sound power levels of machines in the presence of noise from other nearby sources. Narrowband frequency analysis can be used to identify the exact frequency of a tonal component in machinery noise and causally link it with a particular repetitive process occurring within the machine, such as the frequency of a particular fan blade passing or gear meshing. Sophisticated signal-processing methods may also be used for diagnosis, such as correlation, in which two signals are compared to determine the degree of similarity between them. The two signals could be, for example, from the noise radiated from a machine and the vibration of a certain part of the machine, and it is possible to determine the extent to which the noise is produced by the vibration.

2.20 Hearing-Conservation Policies

A hearing-conservation policy is a co-ordinated collection of measures operated by a company aimed at protecting its employees from hearing-loss damage caused by exposure to noise at work. The ingredients of such a policy are:

- Support and commitment to the aims of the policy throughout the company, particularly at senior management level.

- Regular periodic assessments of the noise exposure of employees. In the first instance, this is a statement of the magnitude of the company's hearing-conservation problem, and the basis for the design of the other parts of the programme and for setting priorities. Subsequently, it becomes a check on the progress in implementing other aspects of the programme and the success of its operation. The assessment is more than a survey of noise levels throughout the workplace, although this is an essential part because the process should focus attention on the exposure, and so should include an assessment of the times for which employees are exposed to various levels of noise.

- A programme of measures aimed at reducing the exposure to noise of those employees whose exposure is above some criteria based upon the risk of hearing damage. The criteria used will probably be the action levels of the *Noise at Work Regulations*, although some companies will have their own, more stringent, exposure limits. The noise exposure of employees may be reduced by one of the following methods:

  - Reducing the noise levels in the workplace;

  - Replacing noisy machines and processes by quieter ones;

  - Implementing remedial noise-reduction measures to existing machinery and processes;

  - Separating the employees from the source of noise using automation and employee refuges;

  - Reducing the exposure times (e.g., by job rotation);

  - Using personal ear protectors (i.e., ear muffs or ear plugs);

  - A programme of information and training for all employees in support of all aspects of the policy and aimed at ensuring that everyone is aware of the goals of the policy and of the part they have in ensuring its success;

- Continuing management of all aspects of the policy, including a continuing programme of assessment and reduction of noise exposure; and

- A programme of monitoring audiometry. This is the ultimate check that the other parts of the programme are effective, and usually consists of a pre-employment audiometric check for all new employees and regular follow-up tests, perhaps once or twice each year.

Further information about hearing-conservation policies and health surveillance (i.e., audiometric monitoring) is given in the HSE document Controlling Noise at Work L108 under *the Control of Noise at Work Regulations 2005, Guidance on the Regulations.*

2.21 Appendix 1: The Health and Safety Executive Noise Reduction Topic Inspection Packs: Table E1 on Generic Noise Control Measures

The HSE has produced *Topic Inspection Packs* on noise reduction for various sectors of industry, each in tabular form, summarising the major noise-reduction measures which are appropriate for each sector. This Appendix reproduces the first of these, Table E1, on generic noise control measures, which considers four generic noise-producing activities: air movement; conveying/transporting; forming; processing. The contents of Table E1 are reproduced in **Section 2.21.1** with permission from the HSE.

2.21.1 Generic Noise-Control Measures

2.21.1.1 Activity: Air Movement

Static plant (e.g., compressors, vacuum pumps, blowers) must be relocated/segregated to lesser or non-occupied rooms. The process could be acoustically enclosed within an accessible and adequately ventilated noise-reducing enclosure. Low-noise emission portable generator sets and compressors (e.g., 'hush packs') should be used. The inlet/discharge of fans should be fitted with flexible connections and silencers to reduce duct-borne noise. Excessive line air pressure should be reduced, low-velocity (quiet) nozzles should be fitted to 'open ended' fixed positions, or portable blow-off pipes fitted to remove swarf cuttings, wood chips, or lubricants. Impulsive noise emission from the exhaust ports of pneumatic actuator/manifolds should be reduced using porous metal or plastic port silencers; good connector seals should be maintained to avoid noisy air leaks.

2.21.1.2 Conveying/Transporting

Damped/composite materials should be used for rollers. Component guide/sequencing release mechanisms should be used to reduce component impact noise. Adequate lubrication of bearings/rollers should be maintained. The speed of conveying should be reduced. The conveying ductwork should be suspended using anti-vibration hangers to reduce structure-borne noise. The external damping of ducting conveying materials should be suspended.

Sectional acoustic tunnel hoods should be fitted over open conveyor lines. External damping compounds or rivet plates of sheet metal should be fitted to lightweight flat surfaces (e.g., non-critical machine panel work, chutes, trolley tables, and conveyor sides) to reduce vibration and noise emission. Material stock feeder tubes (e.g., auto-lathe bar stock) should be lined internally.

2.21.1.3 Forming

Machinery (e.g., presses, moulding machines, corrugating machines, bowl polishers, blast chillers or freezers, block making machines, granulators, static compressors, and blowers) should be relocated or segregated to lesser or non-occupied rooms. Machinery could be acoustically enclosed within an accessible and adequately ventilated enclosure. Hydraulic rams should be used to realign distorted fabrications after welding, forming or, alternatively, magnetic damping mats or sandbags can be employed if realigning by hammering. Noise emission from cutting thin sheet metal (e.g., motor vehicle panel work) should be reduced using magnetic damping mats, sandbags and, if feasible, noise can be eliminated by laser-type profiling.

2.21.1.4 Processing

Noisy machinery and/or ancillary equipment (e.g., compressors, presses, fans, saws, cutting-off moulding, fabrication, grinding, and fettling) should be relocated/segregated to lesser and or non-occupied rooms. Machinery could be acoustically enclosed within an accessible and adequately ventilated noise-reducing enclosure. If it is not practicable to remove or enclose, then, for example, long process lines or local noise refuges can be installed for operators to control/oversee processes. If possible, the use of handheld grinders should be minimised by improved component design (e.g., machine weld preparation and removal) or by using 'low noise' discs fitted to portable and possibly pedestal grinders. The sharpness of cutting tools should be maintained and/or reduce speeds with an increased number of cutting teeth or blades.

Falling of cut materials from excessive heights into undamped collection bins can be avoided by use of damped or deadened steel chutes, hoppers, or bins. Materials being cut along their length should be clamped to minimise vibration (i.e., 'bouncing' on supporting surfaces to reduce noise emission from infeed/discharge chutes and hoppers). Damped percussive and rotary percussive tools (e.g., chisels in chipping hammers, rock drills and breakers) should be used. Materials can be broken using quieter hydraulic crushing or bursting rather than percussive methods; for example, crushing concrete instead of using pneumatic or hydraulic breakers; cutting using damped wall saws or diamond wire to profile areas; using bursting methods to remove materials.

2.22 Appendix 2: Summary of Workplace Noise-Control Measures

Another useful summary of workplace noise-control measures is given in the HSE document Controlling Noise at Work L108 under *the Control of Noise at Work Regulations 2005, Guidance on the Regulations*.

2.22.1 Reduction of Noise Exposure by Organisational Control

The various ways in which noise exposure may be reduced by organisational control measures include:

- Plan and organise the work to reduce noise exposure

- Job design

- Job rotation

- Different ways of working:

  - Change of process

  - Change of machine

  - Change of activity

- Workplace design to minimise noise

- Purchasing policy for new quieter tools and machinery in future

2.22.2 Reduction of Noise Exposure by Technical/Engineering Control

The various ways in which noise exposure may be reduced by technical and engineering control measures include:

- Maximise distance between the source and receiver

- Ensure good maintenance

- Minimise air flow (turbulence) noise

- Avoid impacts

- Machine enclosures

- Screens and barriers

- Noise refuges

- Standard noise-control methods:
  - Damping
  - Isolation
  - Use of silencers
  - Active noise control

2.23 Appendix 3: Top 10 Noise-Control Methods

Summaries of the most useful noise-control methods have been produced by Peter Wilson of INVC, and are available on the INVC website (*http://www.invc.com*). Similar versions are also available on the HSE website (*www.hse.gov.uk/pubns/ top10noise.pdf*).

The top 10 solutions are:

1. Damping.

2. Correct fan-installation methods to minimise fan noise.

3. Ductwork – use of sound-absorbent linings.

4. Reduction of fan speed.

5. Use of pneumatic exhaust silencers.

6. Use of pneumatic nozzles using air entrainment.

7. Use of vibration isolation.

8. Using existing machine guards to reduce machine noise by reducing air gaps and linings with sound-absorbing material.

9. Replacing chain drives by quieter belt drives.

10. Replacing existing electric motors with quieter models.

2.22 Summary

A summary of the main point in this chapter is given below:

1. Much can be done to minimise or avoid noise problems without resorting to technical methods of noise control by good planning, management and housekeeping.

2. All noise is generated by vibrating surfaces, impacts, aerodynamic or hydrodynamic forces. In each of these three cases, it is possible to state some simple general principles which can be used to minimise noise production.

3. Noise travels from the source to the receiver by several paths involving airborne or structure-borne sound transmission. Hence, if the noise is not treated at the source it may be necessary, following a priority diagnosis, to apply noise-control treatment to all of the sound-transmission paths.

4. The four principal methods of reducing noise transmission between the source and receiver use sound insulation and sound absorption (for airborne sound) and vibration isolation and damping (for structure-borne sound).

5. Standard-noise control solutions for reducing sound transmission using these four methods are: anti-vibration mounts, screens, enclosures, havens, attenuators and close shielding or acoustic lagging.

6. It is important to select and use the most appropriate materials for controlling noise. This strategy involves understanding and using the correct terminology and the performance specification of noise-control materials.

7. The control of machinery noise at the source involves understanding how the noise-producing forces are being generated by the machine, and how these forces are transmitted to surfaces such as panels, covers and guards, which radiate the noise. This understanding can then lead to methods of reducing the forces or of preventing their transmission to the noise-radiating surfaces.

8. Methods of noise generation and control are considered for some industrial noise sources: fans and blowers, jets and exhausts, punch presses and mechanical handling of materials.

9. Active noise control involves the use of loudspeakers to produce out-of-phase acoustic signals which 'cancel' noise according to the principle of destructive interference. This method works best for low frequencies and has been used to reduce the noise from fans, pumps, and communication headsets.

10. Managers should specify the noise output of any new machinery. Codes and standards on the emission of noise from machinery are available to inform the drafting of such specifications. Purchasers and suppliers of machinery must liaise to ensure that new machinery, when installed, does not lead to unforeseen noise problems.

11. If there are several different sources of noise, then one must establish priorities for noise control. This requires the diagnosis of the most important noise sources and transmission paths.

12. Hearing-conservation policies involve: carrying out surveys of noise levels and assessments of exposure of employees; implementing a programme of measures to reduce noise; issuing of ear protectors to employees supported by a programme of information, instruction, training and supervision to ensure that the protectors are used appropriately; specification of noise limits for new machinery introduced into the workplace; use of audiometric monitoring; effective management and review of all aspects of the policy. The commitment of senior management is important for the success of such policies.

Bibliography

1. I.J. Sharland in *Woods Practical Guide to Noise Control*, Woods of Colchester Ltd., (Flakt Woods), Colchester, UK, 1972.

2. *Noise Control in Industry*, Ed., J.D. Webb, Sound Research Laboratories Ltd., Altrincham, UK, 1976.

3. *Noise Control in Mechanical Services*, Sound Research Laboratories Ltd., Altrincham, UK, 1976

4. R.J. Peters in *Noise Control (A Pira Environmental Guide)*, Pira, Leatherhead, UK, 2000.

5. R.J. Peters, B.J. Smith and M. Hollins in *Acoustics and Noise Control*, 3rd Edition, Prentice Hall, Upper Saddle River, NJ, USA, 2011.

6. D.A. Bies and C.H. Hansen in *Engineering Noise Control*, 2nd Edition, E & FN Spon, London, UK, 1996.

7. BS EN ISO 11690-1:1997 Acoustics – Recommended practice for the design of low-noise workplaces containing machinery – Part 1: Noise control strategies

8. BS EN ISO 11690-2:1997 Acoustics – Recommended practice for the design of low-noise workplaces containing machinery – Part 2: Noise control measures

9. BS EN ISO 11690-3:1999 Acoustics – Recommended practice for the design of low-noise workplaces containing machinery – Part 3: Sound propagation and noise prediction in workrooms

10. BS EN ISO/TR 11688-1:2001 Acoustics – Recommended practice for the design of low-noise machinery and equipment – Part 1: Planning (ISO/TR 11688-1:1995)

11. BS EN ISO/TR 11688-2:2001 Acoustics – Recommended practice for the design of low-noise machinery and equipment – Part 2: Introduction to the physics of low-noise design.

12. Controlling Noise at Work L108 under *the Control of Noise at Work Regulations 2005*, Guidance on the Regulations, Health and Safety Executive, Bootle, Merseyside, UK, 2005.

13. ISO 15665:2003 Acoustics – Acoustic insulation for pipes, valves and flanges, International Organization for Standardization, Geneva, Switzerland, 2003.

14. *Noise from Pneumatic Systems: Guidance Note PM 56*, Health and Safety Executive, Bootle, Merseyside, UK, 1985.

15. *Protection of Hearing in the Paper and Board Industry*, Health and Safety Executive, Bootle, Merseyside, UK, 1988.

16. R.J. Peters, B.J. Smith and M. Hollins in *Acoustics and Noise Control*, 3rd Edition, Prentice Hall, Upper Saddle River, NJ, USA, 2011

3 Noise Control in the Plastics Processing Industries

3.1 Introduction

This chapter reviews the methods that have been used to reduce noise from plastics processing machinery. It re-issues material contained in the *Noise in the Plastics Processing Industry: A Practical Guide to Reducing Noise from Existing Plant and Machinery*, published by RAPRA (Shawbury, UK) in 1985, but also contains some more recent material, mainly from Health and Safety Executive (HSE) publications.

Some of the major sources of noise in the plastics industry are:

1. Impacts – meshing of gear teeth, pellets striking walls, rotation cutters striking plastic load, and saw noise.

2. Out-of-balance forces – vibratory feeders, unbalanced rotating fans, motors, uneven loading in rotating equipment.

3. Aerodynamic – fan and blower noise, compressed air blow-off noise.

4. Hydraulic, pressure pulsations, cavitation, and the formation and collapse of bubbles of gas or vapour within the fluid.

5. Stick-slip friction – squeal from metal-cutting tools, brake squeal.

All these effects have one thing in common: they produce alternating forces within the machine or process.

3.2 Noise Control at the Source

Some methods of reducing noise control at the source include:

1. Can the process be substituted by a different and quieter process?

2. Specify quieter machinery when buying a new plant and equipment.

3. Use resilient inserts to reduce metal-to-metal impacts.

4. Avoid sudden accelerations – change cam profiles.

5. Reduce gear-meshing noises.

6. Substitute softer but better-wearing materials for metal.

7. Line chutes with resilient materials to minimise impact noises.

8. Balance non-essential out-of-balance forces.

9. Isolate vibrating surfaces from adjacent surfaces.

10. Use add-on damping compounds or highly damped materials to prevent ringing.

11. Attenuate compressed air/blow noises.

12. Design for good inlet/discharge conditions to fans and attenuate noise.

13. Observe manufacturer's installation and assembly recommendations for hydraulic systems.

14. Good and regular maintenance.

Methods of noise control during the sound-transmission path include:

1. Increase distance between the source and receiver.

2. Enclose the noise source.

3. Reflect/scatter the source noise with a barrier in direct path between the source to receiver (screen or partial enclosure).

4. Reduce the reverberant field by adding sound absorption to walls/roof.

5. Lag noisy pipes/ducts.

6. Use attenuators to reduce duct-borne noise.

7. Provide noise refuge.

Action plans for noise control should include the following elements:

1. Arrange a preliminary survey to identify noise problem areas.

2. Issue hearing protection as a temporary measure.

3. Arrange a comprehensive survey with noise measurements to establish levels and noise sources.

4. Organise a training programme for all employees and management.

5. Institute a noise-reduction programme by engineering controls.

6. State a buying policy for new machines.

7. Monitor the effectiveness of the noise-reduction measures.

8. Review and update the action plan.

Figure 3.1 is adapted from the 1985 RAPRA book and summarises the results of a survey into the extent of noise hazard from a survey of 88 plastics processing machines.

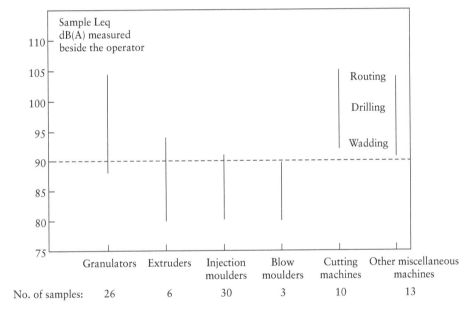

Figure 3.1 Extent of noise hazard from a survey of 88 plastics processing machines. Reproduced with permission from *Noise in the Plastics Processing Industry: A Practical Guide to Reducing Noise from Existing Plant and Machinery*, RAPRA, Shawbury, UK, 1985. ©1985, RAPRA

Tables 3.1 and **3.2** are taken from the 1985 RAPRA book. **Table 3.1** summarises noise sources and noise-control solutions relating to the plastics industry. **Table 3.2** provides a summary of the 25 case studies in the 1985 book. **Table 3.3** presents a more recent (2008) summary of noisy activities and processes, typical noise levels and methods of noise reduction in the plastics industry, and is taken from the HSE website.

Table 3.1 Summary of noise sources and noise control solutions

Machine or process	Noise sources dominating the hazard	Primary noise treatment
Granulating machine (for plastic objects other than sheets)	Interaction of rotating cutters with plastic load. Noise escapes chiefly *via* feed chutes and sometimes powder/granules delivery chute. Vibration of body panels leads to further noise radiation	Chiefly two important remedial measures required: 1) Enclose the whole machine and then feed in material *via* 2) A sound-reducing feed chute; it is vibration-dampened, has sound-resisting barriers inside, and is lined internally with sound-absorptive material
Crumbing machine (for plastic sheets)	Interaction of paddle-type cutter blades with sheet: vibration is from an imbalance of the rotating load. Noise is radiated by machine walls and also escapes through feed and delivery openings. Loose attachments vibrate violently	Ensure that the machine has a sound-reducing lid that can be closed during most of the crumbing cycle. Vibration-damp the external walls of machine or totally enclose. Tighten-up loose attachments (e.g., bolts)
Injection-moulding machines	Hydraulics: pump vane whine, cavitation and pressure pulsations. Motor noise and (in some cases) associated fan noise. Occasionally, pneumatic blow-off noise	• Hydraulics – see separate entry for source reduction. Partially enclose the noisy motor/hydraulics end of the machine to screen operators from this noise • Reduce blow-off pressure and application time • During machine overhaul, replacement of a noisy AC motor with a smaller DC type may achieve further noise reduction
Continuous-extrusion machines	Feed motors and cooling fans	Motor and fan enclosures/silencers or partial enclosures. Replace a noisy AC motor with a DC type during overhaul. Minimise numbers or meshing gears using direct and belt drives
Cut-off saws (frequently associated with extruders)	Interaction of rotating blade with plastic being cut. Also 'ring' of the saw blade	Damp saw against 'ring' with proprietary stick-on material or use a composite saw. Enclose the saw-cutting area as far as reasonable practicable

	Source of noise	Recommendation
Bandsaws	Interaction of travelling blade with plastic being cut	If regular-sized work is cut, then consider enclosure of the machine, otherwise segregate the machine and operator from quieter work areas
Blow-moulding machines	Pneumatic jets and exhausts	Fit silencers to exhausts; reduce blow line pressure; use proprietary, quiet entrainment nozzles. Partial enclosure or noise screening could be employed
Pellet feed systems	Pellets or granules strike pipe walls and may also induce pipe wall 'ring'	Minimise tight bends; vibration-dampen and suitably lag the pipe
Resin pumps	Motor noise and some vibration transmission through pipes	Enclose, vibration-isolate the connecting pipes
Cooling fans	Air turbulence and associated blade passage tones and motor whine	Enclose and provide silenced air inlet and/or outlet
Pneumatic exhausts	Rapid, turbulent escape of compressed air through an orifice	Pipe exhausts away to discharge remotely and safely from operators, or combine exhaust discharges in a common, silenced manifold, or fit adequate and suitable silencers
Ultrasonic welding	Audible screech generated by a vibrating horn and especially the workplace	• Enclose, with sound-reducing access interlocking with the machine controls • In exceptional cases where enclosure is not reasonably practicable, screen the operator from the noise (e.g., with polycarbonate of more than 4 mm)

N.B. This list is not exhaustive

AC: Alternating current

DC: Direct current

Reproduced with permission from *Noise in the Plastics Processing Industry: A Practical Guide to Reducing Noise from Existing Plant and Machinery*, RAPRA, Shawbury, UK, 1985. ©1985, RAPRA

Table 3.2 Summary of Smithers Rapra 25 Case Studies (1985)

	Machine type	Untreated noise level	Noise-reduction treatment	Noise reduction	Cost (1985 prices)	Notes
1	Plastics granulator	96 to 100 dBA	A multi-stage treatment with: sound-insulating cladding to machine surfaces to reduce noise from vibration of machine surfaces; 16-gauge sheet steel cover for cutter blocks with internal sound-absorbing material and well-sealed with rubber edging strips; 16-gauge sheet steel absorbent-lined inlet and outlet tunnels for ventilation openings. Flexible overlapping acoustic curtain strips to reduce noise from work feed inlet aperture. The option of an enclosure was rejected because of a lack of available floor space and anticipated internal cleaning difficulties	15 dBA	£70 + labour (20 man h)	DIY
2	Plastics granulator treatment 101 dBA	85 to 120 dBA depending on the type of plastic and method of processing	Acoustics absorption and baffles in the throat of the machine + entire machine covered in mastic damping compound + acoustic box contains product bin	11 dBA	£120	DIY 100 practical examples (HSE)
3	Large plastics granulator	102 dBA at 20 ft from machine	Commercial enclosure (material not specified) including acoustic duct to feed hopper, acoustic lagging to chip collection cyclone + silencer to air exhaust, with personal access door + light inside enclosure + removable panels for access to feed hopper + silenced ventilation	10 to 20 dBA depending on position	£4,500	Purpose-built commercial enclosure (ICI Acoustics)
4	Substantial enclosure for a plastics granulating machine	–	Double brick-thickness enclosure with steel feed chute (12 gauge lined with 3-inch mineral fibre sound absorbing) designed to give maximum noise reduction and be long lasting plus other measures	30 dBA	£4,000	DIY 100 practical examples

No.	Item		Description		Cost	Notes
5	Conveyor feed to granulator	100 dBA	Conveyor belt removes operator from high levels of noise at the feed aperture	10 to 15 dBA	£3,000	–
6	Acoustics hopper for mobile plastics granulator	95 dBA	Modified hopper cover made of a double-skin 16-gauge steel with 25-mm acoustic absorption cavity	12 dBA	£20	100 practical examples
7	Progressively improved noise-reducing design for granulator	–	Noise-reducing features built into granulator: slower speed, sandwich-steel construction, acoustic cover for feed hopper, built-in flap, anti-vibration mounts	15 dBA	£750	–
8	Enclosure for a pelletiser	–	DIY timber enclosure – 19-mm chipboard with 50-mm absorbent lining secured with pegboard	10 to 12 dBA	£250	–
9	Enclosure for a pelletiser	–	–	25 dBA	–	Commercial enclosure (ICI Acoustics)
10	Retrofit rotors for pelletisers	–	Retro-fit new helical cutters with more cutting blades allows lower speed for same throughput and less noise	12 to 14 dBA	Up to £50,000	–
11	Enclosure for a dicer	–	Brick-built enclosure with 70-cm mineral wool lining. Breakout through the in feed aperture and strip-radiated noise will be limited by forming the aperture as a water-filled bath through which the strip will pass	25 to 35 dBA	£5,000 to £6,000	Commercial enclosure Burgess and Co
12	Enclosure for a dicer	–	Silenced and hinged feed chutes. Access doors for adjustment and maintenance. Facilities for ventilation and services	20 dBA	£3,000	Commercial enclosure (ICI Acoustics)
13	Enclosure for ultrasonic welding machine	96 dBA 8 h LAeq	Steel enclosure with 50-mm mineral wool lining, polycarbonate transparent window with lighting, ventilation	20 dBA	£4,000	–

14	Purpose-designed acoustic cabinet for ultrasonic welding machine	—	Acoustic ducts for convection cooling. Pneumatically controlled vertical sliding window for feed in and a side inspection window for access and maintenance	25 to 30 dBA	£700 to £1,000	—
15	Adhesive tape-slitting machine	—	Control of 'peel off' noise by addition of a supplementary support roller designed to support the tape sheet as close as possible to the peel-off point	4 to 6 dBA	£650	—
16	Inverted closure sensor	—	Sensor based on small jets replaced by a mechanical system	4 to 6 dBA	Up to £900 recouped eventually through compressed air saving	—
17	Silencing of pneumatic exhausts on blow moulding machines	Peak levels of 120 dB	—	10 to 15 dB peak	£35 to £35	—
18	Exhaust plenum chambers	—	Plenum chamber for pneumatic exhausts – an alternative to fitting individual exhaust silencers	20 to 30 dBA	£250 to £300	—
19	Plastic mould Cleaning gun	105 dBA Maximum level	Use of aspirated venture air-entrainment jet	10 dBA max 3 dB 8 h LAeq	Minimal cost plus saving in compressed air	—
20	Noise-reducing modifications to plastics extruder	—	Replace DC motors by more efficient and quieter AC motors. Replace noisy toothed belts with quieter V belts and direct shaft drives	25 dBA	£100	—

21	–	Resonance of cyclone hopper and feed duct feeding an extruder and pelletiser	Resonance reduced by increased damping. Replace square cross-section duct by a stiffer round-section duct	14 dBA	£450	–
22	–	Acoustic lagging of a pipe used to pneumatically convey plastic granules from dicer to hopper	Lagging of 20-gauge steel lined with 40-mm mineral fibre	10 to 15 dBA	£500	–
23	–	Cyclone close shielding and damping	Use of 22-gauge steel with mineral fibre inner lining	10 dBA	£100	–
24	96 dBA	Material transfer by blower replaced by vacuum transfer	Change of process	6 dBA	–	–
25	–	Rotary blowers used to transfer plastic materials	Acoustic box enclosure	7 dBA	£200	–
			Air filter/silencer	4 dBA	£50	

Reproduced with permission from *Noise in the Plastics Processing Industry: A Practical Guide to Reducing Noise from Existing Plant and Machinery*, RAPRA, Shawbury, UK, 1985. ©1985, RAPRA

Table 3.3 Established noise control methods for high-risk additives

High-risk activity/ process	Example noise levels (dB)	Established noise control methods	Further information
Granulators (and other size-reduction machines, e.g., agglomerators, crumbers, shredders, pelletisers)	100 (granulators)	Methods include: • Use feed conveyor to remove operator from higher noise areas • Situate size-reduction machines in separate rooms of buildings – provide for remote or automated feeding • Lag or damp the machine casing • Form a sound trap in feed aperture or hopper • Enclose the machine • Fit segmental or helical cutters • Use a tangential feed • Fit resilient backing to knives • Reduce rotor speed	Example: rubber granulator (see HSE noise Internet pages) Example: enclosure for rubber grinding machine http://www.casestudy.mom.gov.sg/casestudy/case-study-detail.jsp?id=199 Example: strand pelletisers (HSG138 #39) (see HSE noise Internet pages) Example: enclosure for pelletiser http://www.casestudy.mom.gov.sg/casestudy/case-study-detail.jsp?id=294 Generally useful: *Noise in the plastics processing Industry* (RAPRA 1985). This publication is out-of-print but can be obtained by contacting Smithers Rapra, Shawbury, UK – http://www.smithersrapra.com/publications/books
Injection-moulding machines	97–100	Methods include: • Use slow speed pumps • Control release of exhaust air • Mount pumps and motors on anti-vibration mounts and incorporate flexibles hoses in pipelines • Enclose hydraulic power packs • Convert injector guards to acoustic guards • Fit low-noise nozzles to blow guns	Example: controlling release of exhaust air http://www.casestudy.mom.gov.sg/casestudy/case-study-detail.jsp?id=200

Extruders	90	Methods include: • Specify low-noise design • For hydraulic systems, see injection-moulding machines above • Fit silencers to drive motor intakes and exhausts • Enclose drive motor	–
Mould-cleaning guns	105	Replace nozzles with low-noise types (e.g., those which generate an induced secondary air flow). Reduction of less than 10 dB	Example: reduced noise-form mould-cleaning gun (HSG138 #16) (see HSE noise Internet pages)

Figure 3.2 A modular approach to enclosing a noisy machine. Reproduced with permission from P. Mattsson, Rapid Granulator AB, Bredaryd, Sweden. ©Rapid Granulator AB

Figure 3.2 shows an example of a modular type of enclosure, made from a framework of plastic sound-insulating panels that can be adapted for many different noisy machines, including plastics granulators.

3.3 Noise Test Code for Granulators

Information about noise-generating mechanisms and noise-reduction measures is given in *Noise Test Code for Granulators*, which is included as Annex A of BS EN 12012-1:2007+A1:2008: Plastics and rubber machines – size reduction machines.

Section 5.3.2 of BS EN 12012-1:2007+A1:2008: gives the main sources of noise and noise-reduction measures; and states that the main sources of noise are the cutting chamber, hopper, feed opening, discharge opening, suction systems and discharge pipes, if provided.

Among the measures which may be taken are: changing the geometry of the blades, rotor, or hopper; increasing the sound insulation of the cutting chamber; reduction of cutting speed; acoustic enclosures.

More information about noise test codes can be found in **Chapter 8**.

3.4 Conclusion

It might be concluded that, in terms of knowing what needs to be done to make quieter granulators, not much has changed since 1985.

Bibliography

1.   *Noise in the Plastics Processing Industry: A Practical Guide to Reducing Noise from Existing Plant and Machinery*, RAPRA, Shawbury, UK, 1985.

2.   BS EN 12012-1:2007+A1:2008: Plastics and rubber machines – size reduction machines.

3.   *Topic Inspection Pack: Noise*, Health and Safety Executive, Bootle, Merseyside, UK, 2008, p.4.

4 Noise in the Workplace

4.1 Introduction

This chapter discusses the nature of noise-induced hearing loss (NIHL) in the context of noise-exposure levels in the workplace.

The chapter starts with a brief description of how the human hearing system works and of different forms of hearing loss. This is followed by an explanation of how hearing can be damaged by over-exposure to high levels of noise, and how hearing sensitivity is measured.

The requirements of the Health and Safety Executives' (HSE) *the Control of Noise at Work Regulations 2005* are discussed in terms of noise-exposure action levels and limits which, if exceeded, impose duties on employers and employees. Duties of manufacturers and suppliers of noisy equipment are also described.

The first requirement of the regulations is that a risk assessment be carried out to estimate the noise-exposure levels of employees in the workplace and so determine whether their exposure levels exceed the action levels specified in the regulations. A noise-exposure level involves a combination of the level of noise a person is exposed to, and the duration of that exposure. Methods for determining noise-exposure levels in workplaces are described.

Workplace noise-exposure assessments should then be used by employers to develop a 'hearing-conservation action plan', and these are also discussed.

The need for provision of information, instruction and training of employees about hearing-damage risk from noise exposure and what they should do to minimise them are described. The need for good record-keeping with regards to all aspects of hearing conservation are explained and emphasised.

The nature and effects on people of noise-induced hearing is discussed and some information is presented about the numbers of people affected as well as the estimated costs and benefits of hearing-conservation programmes.

The chapter ends with a brief history of the harmful effects of noise exposure in the workplace, which was first recorded 1737 in the cases of industrial deafness among coppersmiths in Italy.

This chapter has used material contained within the HSE Guidance on the Regulations, Controlling Noise at Work L108 under *the Control of Noise at Work Regulations 2005*.

4.2 The Ear and Hearing – The Ear and How it Works

The hearing mechanism consists of three parts; the outer, middle and inner ear (**Figure 4.1**). The outer ear collects the airborne sound waves and directs them down the ear canal to the ear drum, which is a thin membrane or diaphragm which, acting rather like a drum skin, is set into vibration by the sound. The middle ear consists of three little bones: the hammer, anvil and stirrup (often referred to by their Latin names: *malleus*, *incus* and *stapes*) – and collectively known as the 'ossicles'.

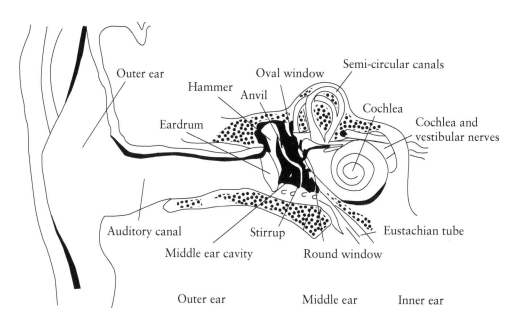

Figure 4.1 Main features of the human ear. Reproduced with permission from *Noise in the Plastics Processing Industry: A Practical Guide to Reducing Noise from Existing Plant and Machinery*, RAPRA, Shawbury, UK, 1985.
©1985, RAPRA

The ossicles, which are connected together, vibrate to and fro. Their function is to transmit the vibration of the ear drum to the inner ear, which is a fluid-filled tube called the 'cochlea', embedded in a cavity in the skull. The malleus is connected to the ear drum and the stapes to the cochlea.

The ossicles themselves are surrounded by air and situated at the head of the Eustachian tube which connects to the back of the throat, so that both sides of the ear drum are always open to the atmosphere. This ensures that the ear drum is not affected by changes in atmospheric pressure, which are much larger than the very small pressure fluctuations caused by the sound waves which the ear is required to detect. It is the temporary failure of the Eustachian tube to equalise the pressure across the ear drum, remedied by swallowing or by sucking a boiled sweet, that we sometimes experience as a pressure on our ears during a change of altitude (for example, take-off during an air flight or going underground into a tunnel). The Eustachian tube may also become blocked by the effects of cold or other infections. Also part of the middle ear is a set of small muscles, called the 'stapediary reflex', connected to the ossicles and which comes into operation very quickly to protect the delicate hearing mechanism from the effects of sudden intense sounds such as gunfire or explosions.

An important function of the ossicles is to act as a bridge between the air in the ear canal (in which the vibrating air particles may have a relatively large amplitude of movement but do not exert much force or pressure) and the fluid in the cochlea (in which the movement of the fluid particles may be much less but causes much larger pressure changes). Without the buffering effect of the middle ear, most of the sound energy striking the ear drum would be reflected rather than being transmitted into the cochlea, as required, rather like the way in which sound reflects off the surface of water in a swimming pool.

The stapes is connected to another diaphragm called the 'oval window', which transmits the vibration to the inner ear. The cochlea is a fluid-filled tube coiled rather like a snail shell, and about the size of a pea. Running through its entire length is a complex structure of membranes and nerve cells, including the basilar membranes and the Organ of Corti that divides the cross-section of the cochlea into two parts. It is in the cochlea that the vibration in the cochlear fluid, caused by the sound waves at the ear drum, is converted, through the action of thousands of tiny 'hair cells' that are stimulated by the vibration, into nerve impulses. It is damage to these hair cells caused by excessive exposure to noise that is the cause of NIHL.

The cochlea is also connected to a structure of three semi-circular canals which controls the sense of balance, and which explains why certain disorders of the ear can affect balance. The nerve impulses generated by the hair cells in the cochlea are transmitted *via* the auditory nerve (which is connected to the cochlea) to the brain, where they produce the sensation of hearing. The vibration that enters the cochlea *via* the oval window travels as a wave along the length of the cochlea to its tip, and back through the second part of the cross-section to another membrane, called the 'round window' which, at its base, communicates with the Eustachian tube and so acts as a pressure-release device.

Some sound energy is also transmitted to the cochlea by 'bone conduction' directly *via* the skull as well as *via* the middle ear.

4.3 Audiometry – The Measurement of Hearing

Audiometry is the science of the measurement of hearing sensitivity. The quantity which is measured is a person's threshold of hearing: the minimum sound pressure level(s) (L_p) which is just audible to that person under defined listening conditions. This varies with the frequency of the sound, and is indicative of the condition of the person's hearing. The baseline against which such measurements are compared is a standardised normal hearing threshold derived from threshold measurements of samples of 'otologically normal' young people (18–25 years).

There are several audiometric test methods but one of the most commonly used is 'pure-tone audiometry'. In the latter, the subject is presented with pure tones, at frequencies of 500 Hz, 1, 2, 3, 4 and 6 kHz, of known sound pressure levels, *via* a standardised and calibrated headset according to procedures defined in ISO 8253-1:2010. The results of such a hearing test are presented as an 'audiogram': a graph of hearing level against frequency. An audiogram is created for each ear, the hearing level being the difference in decibels (dB) between the measured hearing threshold of the person and the normal threshold at the same frequency. Thus, a hearing level of –30 dB means a threshold of hearing 30 dB higher than the normalised threshold, indicating a hearing sensitivity which is 30 dB lower or less sensitive than that represented by the standard threshold. The term 'hearing loss' is also used instead of 'hearing level' although, strictly speaking, hearing loss usually refers to a particular change in hearing level caused by an impairment, or to a deterioration in the hearing threshold of a particular individual.

4.4 Types and Sources of Hearing Loss

There are many forms of hearing loss and a few of the more common types are discussed below briefly. They may be characterised according to whether the condition affects the conductive part of the hearing mechanism (i.e., the outer and middle ear) or the inner ear (i.e., the hair cells in the cochlea) or the auditory nerve and the brain.

The different conditions have different frequency dependencies, and so will produce differing effects on the shape of the audiogram of the subject.

Hearing impairment may also be caused by diseases such as measles, mumps, Meniere's disease, and by the effect of certain (ototoxic) drugs.

Hearing loss can also occur as a result of defects in the transmission of the nerve impulses from the cochlea to the brain *via* the auditory nerve, and as a result of impairment of that part of the brain which interprets these signals ('cortical deafness').

4.4.1 Conductive Hearing Loss

Conductive hearing loss is the result of defects in the conductive parts of the hearing mechanism (i.e., of the outer and middle ear), which prevent sound from reaching the cochlea in the inner ear.

Conductive deafness can be caused by the build-up of wax causing blockage of the ear canal, or by damage to the ear drum or ossicles, which can become fused together so that they are no longer able to vibrate effectively. All of these conditions can be remedied, the blockage simply by removing the wax and the other two conditions by surgery. Infections of the inner ear also affect hearing sensitivity, including 'glue ear' (which causes blockage of the Eustachian tube), which is prevalent among children.

The effect of conductive hearing loss is, in general, like turning-down the amplification of a radio so that the hearing loss is more or less the same across the frequency range.

Otitis media is an infection causing inflammation of the middle ear which can cause the ear drum to be 'sucked in' towards the middle ear through a reduction of pressure in the middle ear.

Otosclerosis is the development of hard or bony deposits at the junction of the ossicles, impairing their usual range of movement.

4.4.2 Sensorineural Hearing Loss

Hearing loss arising from damage to the cochlea is called 'sensorineural deafness' and cannot be remedied. This includes the effects of the natural ageing process, called 'presbycusis', and exposure to high levels of noise.

Unlike conductive hearing loss, the effects of sensorineural hearing loss tend to vary more with frequency, rather like adjusting the individual levels on a graphic equaliser.

4.4.3 Presbycusis

Presbycusis is the loss of hearing sensitivity with increasing age due to the death of hair cells in the cochlea. The process starts with a slight loss of hearing at the higher frequencies, and with increasing age this increases and spreads to lower and lower frequencies, as indicated by the typical audiogram shown in **Figure 4.2**. The degree of hearing loss with age varies from person to person. Quantifying this variability is the subject of the standard BS EN ISO 7029:2000 Acoustics – The statistical distribution of hearing thresholds as a function of age.

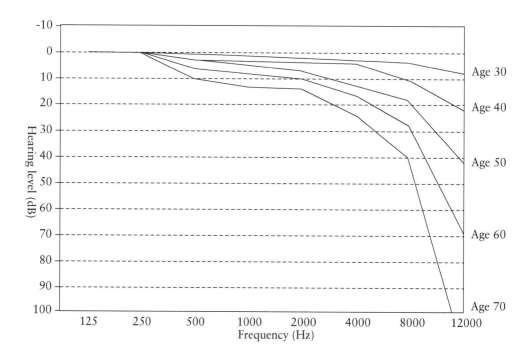

Figure 4.2 Audiograms showing presbycusis. Reproduced with permission from R.J. Peters in *Noise Control (A Pira Environmental Guide)*, Pira, Leatherhead, UK, 2000. ©2000, Pira

4.4.4 Noise-Induced Hearing Loss

NIHL often produces a characteristic shape of audiogram, similar to that shown in **Figure 4.3**. A common feature is a 'dip' or 'notch' in the audiogram showing most hearing loss occurring in the 4 kHz octave band, with this dip increasing in depth and also widening to other frequency bands with increasing duration of exposure. Several explanations for this dip have been given, but a plausible explanation is that the spectrum of much industrial noise arising from noise from machinery often shows the highest levels in the 4 kHz octave band. This is also the frequency range at which the ear is most sensitive to sound. It is also where the consonants that provide much of the information content and which are essential for understanding speech have most of their energy. This often means that a person with NIHL may be aware that someone is talking to them because they can hear the low-frequency vowel sounds but may not be able to understand what is being said because they cannot distinguish the consonants that often occur at the beginning and end of words.

Several good audio demonstrations of the effects of NIHL are available on the Internet.

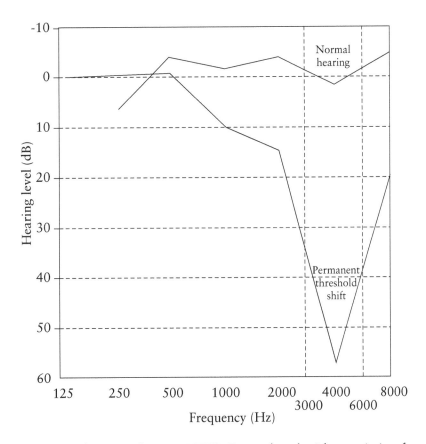

Figure 4.3 Audiogram showing NIHL. Reproduced with permission from R.J. Peters in *Noise Control (A Pira Environmental Guide)*, Pira, Leatherhead, UK, 2000. ©2000, Pira

4.4.5 Acoustic Trauma

Extremely high levels of noise, even of short duration, such as the noise arising from gunfire or from explosions, can cause immediate and permanent hearing damage called 'acoustic trauma'. It is for this reason that the HSE *the Control of Noise at Work Regulations 2005* includes peak action levels that should not be exceeded.

'Tinnitus' is a condition in which noises that are heard by the sufferer are generated in his/her ear rather than originating from any external cause. It is a condition which is experienced by many people in a mild form on an occasional and temporary basis, in connection with a head cold, for example, but which some people experience on a severe and permanent basis. There are several causes of tinnitus that are unconnected with noise exposure. However, tinnitus is also often associated with NIHL: more than 20% of people who suffer from NIHL will also suffer tinnitus.

'Loudness recruitment' is another condition associated with noise-induced hearing damage. In loudness recruitment, there is an abnormal loudness perception above the elevated hearing threshold because the rate of growth of loudness with sound intensity is more rapid than that for those with normal hearing.

4.4.6 Nature and Effects of Noise-Induced Hearing Loss

Several hours of exposure to high levels of noise usually cause an immediate and noticeable temporary loss of hearing, which can be measured as an increase in hearing threshold, and is called a 'temporary threshold shift' (TTS). After a period of relief from the noise exposure the hearing recovers, but continued repetition of the exposure causes the recovery time to increase and the recovery to be less complete, resulting eventually in a noise-induced permanent threshold shift (NIPTS) which is, in effect, NIHL. Recovery times vary, but although substantial recovery can sometimes occur within hours, the process can continue over weeks, depending upon the level, duration and previous history of the exposure, and upon the individual's susceptibility to NIHL. An important practical consequence of this pattern of events is that routine audiometric monitoring of employees at work should ideally be carried out several hours after the last period of noise exposure at work (at least 16 h is recommended), i.e., at the beginning of the working shift, to avoid the audiogram including the effects of TTS.

Evidence suggests that these effects are due to the damage caused to the hair cells and nerve endings in the cochlea. It would seem that the pattern of exposure is significant, with some recovery of the hair cells taking place when exposure ceases, but with permanent damage eventually occurring after repeated prolonged exposure.

The hair cells, once damaged beyond recovery, cannot be regenerated by the cochlea, nor can they be replaced by surgery. The damage to hearing is, thus, permanent and progressive. It is also cumulative, with the effects of NIHL combining with those of presbycusis, which is also due to loss of hair cells but due to age.

4.4.7 Development of a Dose–Response Relationship for Noise-Induced Hearing Loss

The fact that excessive exposure to noise at work causes deafness has been known since the Industrial Revolution. However, it was only in the 1960s that epidemiological studies established a quantitative link between the amount of noise exposure and the level of hearing loss: the dose–response relationship. The link, identified by Burns and Robinson in the UK as well as other scholars, was that it is the amount of A-weighted noise energy received by the ear that is best related to the amount of damage produced. This observation led to the concept of an equivalent (or average) sound (or noise) level in dB over a period of time (LAeq,T). This concept is used to deal with the difficulty of assessing the effects of noise levels that vary throughout the day. This work also led to the publication *Code of Practice Reducing the Exposure of Employed Persons to Noise* published by the Department of Employment (UK) in 1972.

The LAeq,T (also now called the 'average sound level') has remained the basis for estimating the risk of hearing loss due to noise exposure, and the guideline limits of the 1972 Code of Practice eventually became the basis of limits set out in the *Noise at Work Regulations 1989*, which was subsequently replaced by *the Control of Noise at Work Regulations 2005*.

4.4.8 Estimation of the Risk of Hearing Damage

The probability of hearing damage arising from a particular noise-exposure history may be estimated using BS ISO 1999:2013 – Acoustics – Estimation of Noise-inducing Hearing Loss.

This standard enables the probability to be determined of a certain level of hearing damage (for example, 50 dB hearing loss) occurring as a result of a person being exposed to a certain level of noise (for example, 85 dB) over a certain period (for example, 40 years). The standard is used in determining claims for compensation for hearing loss arising from workplace noise exposure. The statistical nature of the dose–response relationship reflects the fact that individuals differ in their susceptibility to NIHL, with wide differences from person to person. Hence, there is still a risk of hearing loss for some people even at exposure levels of below 80 dBA. It is generally considered that the risk becomes negligible below an exposure level of 75 dBA.

4.4.9 Combined Effects of Noise-Induced Hearing Loss and Presbycusis

The combined effects of NIHL and presbycusis affect the hearing levels of high frequencies more than those of lower frequencies. The main social problems resulting from this are to do with understanding speech. This is because much of the information content in speech is carried by the consonants, which contain more of the higher frequencies in speech, particularly sibilants such as 's', 't' and 'p' whereas the vowels, which contribute to the character and quality of the speech, contain more of the lower frequencies. Thus, the listener may have difficulty in distinguishing between words with the same vowel sounds but beginning or ending with different consonants, such as 'car', 'cart', 'carp' as well as 'thought' 'taught', 'sought' and 'bought'. Therefore, the effect of the damage to the hair cells is not simply to reduce the level of the sound that is heard, which would be rather like turning down the volume on the radio, and which could possibly be assisted using hearing aids to boost the volume again, but to remove information (i.e., the high frequencies) from the speech signal as well as lowering its volume, and this cannot be satisfactorily remedied by hearing aids. The result is the very frustrating one that even though the sufferer can hear the speech it appears muffled and indistinct (rather like seeing through frosted glass). The frustration may be aggravated by the effects of tinnitus.

The effects of NIHL are all the more insidious because they occur gradually, and at first may be undetected until considerable irreparable damage has occurred. The sufferer may learn to adapt, sometimes without realising it, by turning up the volume of TV or radio, by sitting closer to the speaker in an audience and by paying more attention, and always turning to face the speaker and watching his or her lips during face-to-face conversations. The most serious problems occur when trying to participate in group conversations, when the sufferer does not always have direct face-to-face contact with the speaker, and when the conversation is taking place against a background of noise, such as those in pubs and at parties, and when using the telephone. When listening to the radio or TV as part of a group, the sufferer may not be able to hear at the levels selected by most of the group, and may risk becoming unpopular if constantly requesting the volume to be increased. Conversations with the very young or with strangers, who do not have either the patience or the understanding to make allowances for the effects of the sufferer's condition, may become difficult.

In extreme cases, these effects can lead to feelings of inadequacy, rejection, loneliness and isolation, and may be compounded by worries that the hearing loss may lead to inability by the sufferer to perform his/her job effectively and hence to further anxiety about job security.

It is because the onset of NIHL is gradual and may not at first be noticeable to the subject that it is important that those exposed to high levels of noise either at work or as a result of noisy leisure activities should undergo regular audiometric tests to check their hearing level.

4.5 Otoacoustic Emission

The ability of the ear to distinguish between different frequencies was first explained by a theory of cochlear mechanics proposed by Bekesy in 1960. Certain difficulties with this theory have prompted research in the past few years that has led to new understanding of the workings of the inner ear, and which may lead to new ways of detecting and measuring NIHL.

These difficulties relate to the ability of the ear to detect and discriminate between different frequencies in a sound. According to the 'place theory' of Bekesy, sounds of different frequency produce maximum amplitudes of vibration and, therefore, maximum stimulation of the hair cells at different positions along the length of the cochlea. Experiments on human frequency selectivity indicate that the cochlea can, in effect, be thought of as being divided into critical frequency bands, corresponding approximately to 1/3 octave bands, each having its own location along the cochlea. Bekesy's theory proposed that the place effect could be explained in terms of the mechanical vibration of the basilar membrane in the cochlea, and its frequency response. However, experiments in which the cochleas from dead animals were subjected to vibration failed to produce the necessary degree of frequency selectivity because the cochlea material was too well damped. This led to revised theories of cochlea vibration in which the basilar membrane is part of an active vibration system involving feedback from the hair cells (rather like the servo-mechanisms in engineering control systems) and which could work only in a living cochlea and be controlled by the brain. These ideas led to experiments in which signals emitted by the cochlea in response to sound striking the ear drum were detected by sensitive miniature microphones placed very close to the ear drum of the test subject. Although these signals, or otoacoustic emissions, have sometimes been called 'cochlear echoes', they are really the active signals emitted by the cochlea in response to sound stimuli.

This new understanding of cochlear dynamics has already found practical application in the testing for deafness in babies and young children through the monitoring of otoacoustic emissions. It is possible that in the future similar methods could be used as an alternative or supplement to audiometric testing in the diagnosis and monitoring of NIHL.

4.6 Summary of the Control of Noise at Work Regulations 2005

These regulations follow the requirements of EC Directive 2003/10/EC – Noise. They replace the HSE *Noise at Work Regulations 1989* which had been in force since 1990.

The regulations define an upper and lower action value and a limit value which determine the duties of the employer and employees. In each case there is a daily (or weekly) noise-exposure level measured in dBA, and a peak sound pressure level measured in dBC

4.6.1 Action and Limit Values

The action and limit values are:

- Lower exposure action value: A personal daily (or weekly) noise-exposure level of 80 dBA, and a peak sound pressure level of 135 dBC.

- Upper exposure action value: A personal daily (or weekly) noise-exposure level of 85 dBA and a peak sound pressure of 137 dBC.

- Exposure limit value: A personal daily (or weekly) noise-exposure level of 87 dBA and a peak sound pressure level of 140 dBC.

The exposure limit value, which must not be exceeded, may take into account the effect of wearing hearing protection.

4.6.2 Duties of Employers

The duties of employers are:

- Assess risks to employees (Regulation 5).

- Take action to reduce noise exposure and ensure legal limits on noise exposure are not exceeded (Regulation 6).

- Provide employees with hearing protection and, where appropriate, create hearing-protection zones (Regulation 7).

- Maintenance and use of equipment (Regulation 8).

- Carry out health surveillance (Regulation 9).

- Provide employees with information, instruction and training (Regulation 10).

4.6.2.1 Assess Risks to Employees

A risk assessment must be carried out if any employee is likely to be exposed to noise at or above the lower exposure action values.

The lower exposure action value is 80 dBA. A noise level of 80 dBA would be intrusive, but it would still be possible to hold a conversation; it is typical of noise levels in a busy street or crowded restaurant, or the noise from a vacuum cleaner. At a noise level of 85 dBA it would be necessary to shout to talk to someone 2 m away, and at a noise level of 90 dBA it would be necessary to shout to talk to someone 1 m away.

4.6.2.2 Take Action to Reduce Noise Exposure

There is a general duty to reduce risk from exposure to noise a far as is reasonably practicable, even in situations where noise-exposure levels are below the lower exposure action value.

If any employee is likely to be exposed to noise at or above the upper exposure action value, the employer should reduce exposure to as low a level as is reasonably practicable by establishing and implementing a programme of organisational and technical measures, excluding the provision of personal hearing protectors.

4.6.2.3 Provide Employees with Hearing Protection

Hearing protection should be provided above the lower exposure action value if requested by employees but, in this situation, the regulations do not make their use compulsory.

If employees are likely to be exposed at or above the upper exposure action values, employers must provide hearing protectors and must ensure the hearing protectors are used, and regulations require employees to use them (Regulation 8).

Hearing-protection zones must be designated in any area of the workplace in which the upper exposure action values would be likely to be exceeded if personnel spent a significant portion of the working day within them, even if access is generally infrequent (e.g., plant rooms or compressor houses). Hearing protection should be worn at all time in a hearing-protection zone.

4.6.2.4 Ensure Legal Limits on Noise Exposure Are Not Exceeded

This refers to exposure levels above the exposure limit value, which includes the attenuation provided by any hearing protection that is being worn.

4.6.2.5 Maintenance and Use of Equipment

Employers should ensure that anything provided under these regulations is maintained in an efficient state, in efficient working order and in good repair, and is fully and appropriately used.

4.6.2.6 Carry Out Health Surveillance if the Risk Assessment Indicates that there is a Risk to Health

Employers should provide regular hearing checks to employees exposed regularly to above the upper exposure action values. If exposure is between the lower and upper exposure action values, or if employees are exposed only occasionally to above the upper exposure action values, regular hearing checks should be provided if there is evidence that an individual may be particularly sensitive to noise.

Employees who may be particularly vulnerable to risks from noise exposure include: those with a pre-existing hearing condition, or with a family history of deafness; pregnant women; young people; those also exposed to high levels of hand-transmitted vibration; those exposed at work to certain ototoxic substances, particularly solvents.

4.6.2.7 Provide Employees with Information, Instruction and Training

Employers should provide employees who are exposed to noise which is likely to be at or above a lower exposure action value with suitable and sufficient information, instruction and training about the risks to hearing from exposure to high levels of noise and what should be done to minimise these risks.

This should include:

- The nature of NIHL and its possible effects on people's lives.

- Likely noise-exposure levels and risk to hearing.

- What is being done to control risks and exposures.

- Where and how employees can obtain hearing protection.

- How to report defects in hearing protection and noise-control equipment.

- Duties of employees under the regulations.

- What employees should do to minimise risk (e.g., how to use, store and maintain hearing protectors and other noise-control equipment).

- What health surveillance system is being provided.

- How to detect the symptoms of hearing loss and how and to whom to report them.

4.6.3 Duties of Employees

The duties of employees are to:

- Comply with measures put in place by employers to reduce noise exposures.

- Inform management if any such measures (including hearing protection) are in need of maintenance or replacement.

- Wear hearing protection if noise-exposure levels are above the upper exposure action values.

Employees, therefore, have a duty to use noise-control measures such as exhaust silencers and machine enclosures as instructed, to remove hearing protectors and all noise-control equipment and to promptly report defects in hearing protectors and noise-control equipment, for example, damage to enclosures, worn seals to doors, and exhaust silencers in need of replacement.

4.7 Duties of Suppliers and Manufacturers of Equipment

Suppliers and manufacturers of equipment have duties under two European Union (EU) directives currently in operation: The EC Directive 2006/42/EC – Machinery Noise and the EC Directive 2000/14/EC – Noise – Equipment for Use Outdoors.

The directive on machinery noise is concerned mainly with noise in the workplace and protecting hearing of employees from damaging levels of noise exposure, and the outdoor directive is related to minimising disturbance from noise from outdoor machinery in the environment.

Both directives require that noise emissions from machinery should be openly declared and that manufacturers should be required to reduce the noise of their products by the use of noise-reduction techniques, as far as possible. Both directives are helpful to purchasers of equipment and machinery by informing them of noise-emission levels from different competing products, and it is the intention that this will be an incentive to manufactures to develop quieter machines to obtain competitive advantage.

In the UK, the machinery directive is implemented in *'The Supply of Machinery (Safety) Regulations 2008'* (Statutory Instrument 2008 No.1597).

The following information relating to noise emissions must be declared in sales literatures, handbooks, and instruction manuals:

- A-weighted emission sound pressure level (L_{pA}) at workstations, where this exceeds 70 dBA; where this level does not exceed 70 dBA, this must be stated.

- Sound power level L_{wA} emitted by the machinery where the L_{pA} at workstations exceeds 80 dBA.

- Peak C-weighted instantaneous sound pressure levels at workstations, where this exceeds 63 Pa (130 dB relative to 20 µPa).

More details are given in **Chapter 8**.

4.8 Carrying Out Risk Assessments

Carrying out a risk assessment requires an estimate to be made of the noise-exposure level of each employee in the workplace with sufficient accuracy to determine whether one or more of the various action values or limit values has been exceeded.

Because of the wide variety of noise-exposure patterns which can exist in different workplaces, there can be no fixed procedure for carrying out such assessments. Any one employee might, for example, work in the same place throughout each day, or may be continually moving from place to place, and in either case may be subject to a more or less steady pattern of noise exposure throughout each day, or to one which is continually changing. Furthermore, the noise-producing activities and noise-exposure patterns of employees in the workplace may remain more or less constant from day-to-day, or might be completely different, and on any one day different employees may have different exposure patterns.

Therefore, the person carrying out the noise-exposure assessment must have, or acquire, a good understanding about what is happening throughout the workplace on the day of the assessment and devise the best approach. Whatever is decided, it will require the gathering of information about noise levels and exposure times at each noisy activity for each employee, and then combining them to form a daily noise-exposure level to feed into the risk assessment.

Sound-level measurements are needed at positions where people work and, furthermore, at employee ear positions.

There are three ways of measuring noise levels: using fixed measurement positions, by taking activity samples, or by using dosemeters.

4.8.1 Fixed Positions

At fixed machine operator positions, this is best done, if possible, with the operator temporarily absent. However, if this is not possible, the microphone should be held about 15 cm from the employee's ear. If employees move around the workplace, sound levels should be measured at several typical locations.

4.8.2 Activity Samples

If employees undertake repetitive routine activities which involve movement over a limited area of activity, then sampling may be carried out. The surveyor/assessor holds a hand-held sound-level meter (with the agreement of the employee) about 7.5 cm to the ear of the employee while a typical repetitive task is carried out, which involves some movement but only over a fairly limited area.

The result will be an average noise level over the activity period, for example, 5 or 10 min. If the employee repeats this activity so many times per hour, or so many times each work shift, then it will be possible to calculate the average noise level at his/her ear due to this activity over the entire working day.

4.8.3 Noise Dosemeters for Mobile Employees

The third method is to fit a dosemeter to the employees. This is a small, pocket-sized portable device which can be worn by the employee and connected by a cable to a microphone secured by a clip to the wearer (often at a lapel) or at shoulder height, and is carried about as he/she moves around the workplace over a prolonged period of the day or for the entire shift. Smaller versions are available for which the entire device, including the microphone, is worn without the need for a cable, as a 'dose-badge'. 'In-ear' dosemeters have also been developed.

4.8.4 Data-Logging Dosemeters

An obvious advantage is that the microphone records the noise exposure of the employee continuously throughout the day as he/she moves around the workplace. Therefore, dosemeters are often considered to be the ideal noise-monitoring solution for employees who move around the workplace. However, they have two main disadvantages. First, the sound-level measurements may not be as accurate as those from either a fixed position (maybe tripod-mounted) or even a hand-held microphone because the dosemeter microphone is so close to the wearer's body and can be affected by sound being reflected, scattered or absorbed by his/her body. Second, even with data-logging dosemeters, the dosemeter microphone will not be visible to the surveyor for most of the time and he/she will not know for certain exactly what the microphone has measured. The dosemeter readings could become contaminated, either deliberately or inadvertently, with 'false' signals, for example, from the microphone impacting with nearby surfaces, from the voice of the wearer, or maybe from the dosemeter being removed temporally from the wearer's body.

Therefore, dosemeter readings should be examined carefully and compared with sound-level readings taken under more controlled conditions.

It is important to select both the locations and timings of all the measurements very carefully so that they are truly 'representative' of the noise-exposure patterns. All sound-level and dosemeter readings should be considered as measurement 'samples' and, if possible, should be repeated to check for variability between samples so that (if necessary) an average value can be obtained. Worst-case scenarios should be considered and measured, but judgement used so that they are not exaggerated in the final analysis (i.e., it should be a 'reasonable' worst-case scenario that evaluates more typical and average conditions).

4.8.5 Workplace Noise-Level Maps

If the workplace noise-exposure pattern follows a more or less fixed routine from day-to-day it can be useful to create a 'noise-level map' of the workplace. This can be achieved by taking average noise-level measurements over a regular grid of positions covering the entire workplace. The various levels can then be plotted on a grid overlaying a floor plan or map of the workplace. If required the map can be used to plot the noise-level contours of the workplace. Commercial software packages are available to plot such contours.

The HSE has produced an information sheet (available online) on noise mapping in paper mills, which could usefully be applied to other industries (HSE information sheet *Noise Mapping in Paper Mills*, Paper and Board Information Sheet No.2).

4.8.5.1 Peak Action Levels

If there are short bursts of very high levels of noise, these should be included in the monitoring of average levels using the methods described above. Separately, if necessary, peak noise levels should be measured to determine whether peak action or limit values have been exceeded.

Examples of processes producing high-peak noise levels include: general mechanical handling (bangs and clatters), impact tools, hammering (general metal working, and 'knocking out' of castings), punch presses, jolt squeezing moulding machines and nail guns.

There is an ISO standard which describes methods for determining workplace noise-exposure levels: ISO 9612:2009 Acoustics – Determination of Occupational Noise Exposure – Engineering method.

4.9 Reporting the Risk Assessment/Noise-Exposure Assessment

4.9.1 Recording the Risk Assessment

The record should contain:

- The scope of the assessment: workplaces, areas, jobs, people and activities included.

- The date(s) that the assessment was made.

- Daily personal exposures (and peak exposure levels) of employees, or groups of employees included.

- The information used to determine noise exposures.

- Details of noise measurements and procedures, including the name of the person responsible.

- Any further information used to evaluate risks.

- The name of the person who made the risk assessment.

- Action plan to control noise risks.

4.9.2 When to Review the Risk Assessment

Review of the rick assessment should be completed when the work patterns, processes, procedures, machinery have changed significantly:

- If new technologies (new ways of working, or noise-control techniques) have become available.

- If new noise-control measures have been introduced.

- If health surveillance shows the need for review.

- If measures previously unjustifiable become reasonably practicable.

- If 2 years has elapsed since the last assessment.

4.10 Action Plans

The purpose of the risk assessment and noise-exposure assessment is not to be filed away but to be the basis of an action plan to improve as far as possible the noise-exposure levels in the workplace and, therefore, to reduce risk of hearing loss arising from noise exposure.

The action plan should contain:

- What has been done to tackle immediate risks.

- Actions being considered regarding the general duty to reduce risks.

- Plans to develop a programme of noise-reduction measures.

- Arrangements for provision of hearing protection and hearing-protection zones.

- Arrangements for providing information, instruction and training.

- Arrangements for providing health surveillance.

- Realistic time scales for the above to be carried out.

- People or post holders to be responsible for the various tasks mentioned above.

- Person with overall responsibility to ensure that the plan is competently carried out.

4.11 Record-Keeping

It is important to keep records of all actions taken to reduce noise-exposure levels in the workplace and to further protect the hearing of employees. Records which might be kept include:

- Noise surveys/risk assessments.

- Actions plans.

- The introduction of new quieter equipment.

- The introduction of noise-control measures.

- The issue of ear protectors.

- Information, instruction and training given to employees.

- Health surveillance records.

- The magnitude of the problem – the numbers of people affected by NIHL and the consequences.

- Meetings regarding all of the above.

4.11.1 Record-Keeping In the UK

The following quotation is from the introduction to the HSE's Guidance on the Regulations, Controlling Noise at Work L108 under *the Control of Noise at Work Regulations 2005*:

> '*Hearing damage caused by exposure to noise at work is permanent and incurable. Research estimates that over 2 million people are exposed to noise levels at work that may be harmful. There are many new cases of people receiving compensation for hearing damage each year, through both civil claims and the Government disability benefit scheme, with considerable costs to industry, society and, most importantly, the people who suffer the disability.*'

A more recent statement was issued by HSE as background to their *Listen Up!* Symposium held on 2nd March 2016:

> *'Over 1 million workers in the UK are exposed to noise above the legal action value and therefore at risk of hearing damage. In addition to an increase in social and leisure noise exposure for younger generations, the increasingly ageing working population means that more workers will exhibit signs of hearing impairment...'*

> *...The UK insurance industry is currently paying £70 million per year in deafness related claims and there has been a substantial increase in the number of claims for NIHL in recent years.'*

4.11.2 Record-Keeping Worldwide

Although it is very difficult to obtain reliable and consistent estimates of the number of people exposed to damaging levels of noise, a review of the literature (between 1995 and 2016) indicates that hundreds of millions of people are exposed to damaging levels of noise.

According to estimates by the World Health Organization (WHO), estimates of the number of people affected worldwide by hearing loss increased from 120 million in 1995 to 250 million worldwide in 2004. Much of this impairment may be caused by exposure to noise at work.

Noise-induced hearing impairment is the most prevalent irreversible occupational hazard according to *WHO Community Noise Guidelines 1999*.

NIHL remains one of the most prominent and most recognised occupational diseases in the Member States of the EU. In the EU, 28% of workers surveyed reported that at least one-fourth of the time they are occupationally exposed to noise loud enough that they would have to raise their voices to hold a conversation (which corresponds to approximately 85–90 dBA).

It has been estimated that between 10 and 20 million workers are exposed to damaging levels of noise in the USA.

The highest attributable fraction of adult-onset hearing loss resulting from noise exposure in the world comes from Asian countries. NIHL is a serious health problem in Asia, not only because of the number of affected labourers, but also because most Asian countries are still developing economies where access to health services and preventive programmes are limited.

4.12 Costs and Benefits of Measures to Reduce Noise in the Workplace

The costs and benefits of the *Noise at Work Regulations*, introduced in 1990, have been estimated in a research project commissioned by HSE Contract Research Report 116/1993: *The Costs and Benefits of the Noise at Work Regulations 1989*. The costs fall mainly on employers, while the benefits accrue mainly to employees and to society as a whole. The provision of hearing protectors was seen by employers in the survey as the most significant element of noise control in terms of cost and effectiveness for risk reduction. Noise assessments were also seen as effective measures but relatively costly. On the whole, employers did not see the cost of noise control as being particularly significant.

It was estimated that the cost to employers of complying with the regulations was almost £50 million in the first year and £37 million per year in the following years. On an annual basis, this represents £35.60 for each worker exposed to noise. The principal costs arose from conducting assessments (£21 million for the initial assessment and almost £10 million annually for subsequent assessments), noise-control measures (about £8.5 million a year), providing ear protection (£9.5 million a year, or £8.60 per noise-exposed employee), and training (£7.3 million a year, or £5.60 per noise-exposed employee). The average cost of ear protectors was £9.65 for muffs and £0.39 for plugs. The cost of buying a quieter plant or of modifying an existing plant to reduce noise levels was between 1 and 5% of the overall budget. Other significant costs identified were those of management time in dealing with compensation claims, higher insurance premiums resulting from such claims, and time spent ensuring compliance with the regulations. The cost of audiometry, undertaken by about 25% of the firms surveyed, was about £8 per employee.

Although in many cases the main reason for introducing noise-control measures was found to be compliance with the regulations (i.e., the avoidance of a penalty), the following possible benefits to employers were also identified:

- Improved efficiency (i.e., a perception that a quieter workplace was more efficient);

- Reduced stress and improved morale;

- Increased productivity;

- Reduced absenteeism;

- Reduced management time spent dealing with deafness-compensation claims;

- Reduced insurance premiums;

- Reduced turnover of staff;

- Raising the profile of health and safety in the workplace;

- Demonstrating a professional approach to clients;

- Demonstrating commitment to quality management processes, such as total quality management; and

- Demonstrating a caring culture within the organisation.

4.13 Hearing-Damage Claims

According to the *Tackling the Compensation Culture: Noise Induced Hearing Loss Claims* article published by the Association of British Insurers (*http://www.abi.org. uk*), the number of claims for compensation for NIHL reached a peak in the early 1990s and has reduced markedly since then, but there has been an increase since 2010 due to the introduction of a 'no win no fee' claims culture in the UK:

> '*Despite the improvements in health and safety measures and better regulation, insurers and compensators have experienced a sharp increase in NIHL claims in recent years resulting from the UK's compensation culture. In 2010, there were 24,352 NIHL claims notified. This increased to 85,155 notified claims in 2013, which represents an increase of almost 250% and these claims had a total estimated cost of over £400 million.*'

This is a very good reason for keeping records relating to all aspects of hearing conservation, as discussed above.

The best defence against unjustified NIHL claims is to have in place a hearing-conservation programme (including pre-employment audiometry) which complies with all the requirements of *the Control of Noise at Work Regulations 2005*, or better, and to keep comprehensive records.

4.14 A Brief History of Noise-Induced Hearing Loss

A brief history of noise-induced hearing loss in the form of a chronology of important events is given in **Table 4.1**.

Table 4.1 A brief history of noise induced hearing loss	
1700	The risk of hearing damage due to exposure to high levels of noise has been known about for a long time. In his classic text *De Morbis Artificum Diatrib'* (Diseases of Workers), first published in 1700, the Italian physician, B. Ramazzini (4th October 1633–5th November 1714) identified deafness in coppersmiths in Venice as being due to exposure to noise arising from their work of hammering copper to make it ductile.
	With the coming of the industrial revolution and steam power, the problem of occupational noise exposure became acute. Workers making steam boilers, in particular, succumbed to deafness in such numbers that it became known as 'boiler-makers' disease.
1831	First authoritative reference to noise as a health problem: Dr. J. Fosbroke, writing in the *Lancet*, stated that 'blacksmiths' deafness is a consequence of employment'.
1886	First quantitative study of NIHL by Dr. T. Barr. He studied the loss of hearing in Glaswegian boiler makers. On average, they could hear quiet sounds at just one-tenth the distance at which those with normal hearing can.
1963	The *Wilson Report*:
	In April 1960, a committee was set up 'to examine the nature, sources and effects of the problem of noise and to advise what further measures can be taken to mitigate it.'
	The final report of the committee on the problem of noise, widely known as the *Wilson Report* after the name of its chairman, was issued 3 years later in March 1963. The report was 234-pages long, contained 14 chapters, and covered all aspects of noise, and its recommendations shaped the approach to noise legislation and mitigation up to the present day. Chapter 8 was on noise from industry and chapter 13 was on occupational exposure to high levels of noise. This chapter recommended that further research should be undertaken, and that 'the Ministry of Labour and other interested government departments should consider whether the time has not then come to lay down by legislation minimum standards to protect workers against damaging noise exposure in industry.'
	Many important initiatives and developments followed in the years following the *Wilson Report*, including the Department of Employment Code of Practice on *Reducing the Noise Exposure of Employed Persons to Noise, 1972* which subsequently, *via* EC directives, became the *Noise at Work Regulations 1989* and then the *Control of Noise at Work Regulations 2005*.
	The UK Ministry of Labour published *Noise and the Worker*, which introduced the concept that excessive noise in the workplace could lead to hearing loss. It was recommended that employees should not be exposed to a noise level of over 90 dB over an 8 h working day. Later, case law held that from 1963 onwards employers should have been aware of the risk of exposure to loud noise at work. Despite the widespread previous recognition of noise-induced deafness, the date of issue of this booklet is generally considered in common law cases as the time from which employers should have been aware of the risk and taken action to protect their employees. This in effect meant that claims for NIHL before this date were unlikely to be compensated.

1972	Publication of Department of Employment (UK) Code of Practice *Guide to the Exposure of Employed Person at Work.*
1974	The Health and Safety at Work Act imposed duties on employers to protect employees against risk of hearing damage as a result of exposure to noise at work.
1983	Publication by HSE of *100 Practical Applications of Noise Reduction Methods.*
1985	Publication by RAPRA of *Noise in the Plastics Processing Industry: A Practical Guide to Reducing Noise from Existing Plant and Machinery.*
1989	The *Noise at Work Regulations 1989* introduce an effective system of 'action levels', involving both the employer and employee in the process of hearing protection.
1995	Publication by HSE of *Sound Solution Case Studies* (60 case studies).
2005	The *Control of Noise in the Workplace Regulations*, which reduced the level of the upper and lower action values by 5 dB, introduced an exposure limit value, and health surveillance measures (audiometry).

4.15 Summary

The main points covered in this chapter are summarised below:

1. Sound is transmitted by the outer and middle ear to the inner ear or cochlea, where it is converted to biological signals transmitted to the brain and interpreted as sound.

2. It is permanent and irreparable damage to the hair cells in the cochlea caused by excessive exposure to noise which is responsible for NIHL.

3. Hearing sensitivity is described in terms of the hearing level, which is the threshold of hearing of the test subject, measured using an audiometer, and compared with a standardised hearing level. Audiograms show the subject's hearing level for each ear at different frequencies plotted as a graph.

4. Continuous exposure to high levels of noise at work causes TTS, NIPTS, loudness recruitment, and can also cause tinnitus.

5. Risk of NIPTS is related to the duration and level of exposure to noise, measured as LAeq,T over an 8 h period.

6. NIPTS, combined with the effects of presbycusis, can cause difficulty in understanding speech, resulting in a severe social handicap for the sufferer.

7. Air conduction audiometry has been briefly described and typical audiograms showing presbycusis and NIHL have been presented.

8. Otoacoustic emission and its applications have been briefly described. In the future, the technique could be used as an alternative or supplement to audiometric testing for the diagnosis and monitoring of NIHL.

9. *the Control of Noise at Work Regulations 2005* set out upper and lower exposure action values and a limit value which determines the duties of the employer and employees. In each case, there is a daily (or weekly) noise-exposure level measured in dBA, and a peak sound pressure level measured in dBC.

10. The nature of peak exposure action levels and when peak sound levels should be measured have been explained.

11. Duties of employers are to: assess risks to employees; take action to reduce noise exposure and make sure legal limits on noise exposure are not exceeded; provide employees with hearing protection and, if appropriate, create hearing-protection zones; ensure maintenance and use of equipment; carry out health surveillance; provide employees with information, instruction and training.

12. Duties of employees are to comply with measures put in place by employers to reduce noise exposures; inform management if any such measures (including hearing protection) are in need of maintenance or replacement; wear hearing protection when noise-exposure levels are above the upper exposure action values.

13. Suppliers and manufacturers of equipment have duties under EC Directive 2006/42/EC to declare noise-emission levels of their plant and equipment.

14. Carrying out risk assessments requires an estimate of employees' noise-exposure levels. This may involve a combination of noise-measurement techniques using sound-level meters (for static and activity sampling) and noise dosemeters. In addition, noise maps of the workplace showing noise level contours can sometimes be useful.

15. Information that should be included in risk assessment reports has been listed together with advice about when risk assessments need to be reviewed.

16. A list of suggested contents for a programme of information instruction and training has been presented.

17. Action plans – the purpose of the risk assessment and noise-exposure assessment is not to be filed away but to be the basis of an action plan to improve as far as possible the noise-exposure levels in the workplace and, therefore, to reduce risk of hearing loss arising from noise exposure. Suggested contents for inclusion in action plans have been presented.

18. It is very important to keep records of all activities related to hearing conservation. Comprehensive records can be used to demonstrate good practice, compliance with regulations and to defend against hearing-damage compensation claims.

19. A brief review of the prevalence of NIHL and of numbers of people exposed to damaging levels of noise has been presented, and of the costs and benefits of hearing-conservation programmes.

20. NIHL at work is one of the major industrial health problems in Europe and the UK, with millions of employees exposed to more than 85 dBA and with large numbers of people receiving industrial injury benefit and being awarded compensation claims.

21. The cost to employers of complying with *the Control of Noise at Work Regulations 2005* has been estimated to be £37 million per annum, or £35.60 for each noise-exposed worker. The major components of this cost are from carrying out assessments, implementing noise-reduction measures (including purchase of new, quieter machinery), issuing hearing protection and training. The benefits to the employer are difficult to quantify, but have been estimated at 25% of the costs. There are further benefits to the individual employee and to society.

22. The attitude of senior management is the critical factor influencing the success of hearing-conservation programmes in a company. Noise-exposure assessments that give an action plan for managers to implement are a key stage in the programme. Although hearing protection is issued widely throughout industry, it is often not being used effectively by employees because of poor training and management supervision.

23. The large number of claims for hearing-damage claims received by industrial insurance companies in the 1960s–1980s has steadily declined since the early 1990s, but has increased again since 2012, as a result of a claims culture of no win, no fee.

24. A brief history of NIHL has been presented.

Bibliography

1. Controlling Noise at Work L108 under *the Control of Noise at Work 2005 Regulations*, Guidance on the Regulations, Health and Safety Executive, Bootle, Merseyside, UK, 2005.

2. BS ISO 1999: 2013 – Estimation of noise induced hearing loss.

3. ISO 9612:2009 Acoustics – Determination of occupational noise exposure – Engineering Method.

4. W.J. Murphy, *Acoustics Today*, 2016, **12**, 1, 28.

5. D. Nelson, *American Journal of Industrial Medicine*, 2005, **48**, 6, 446.

6. *Noise in Figures*, Risk Observatory, Thematic Report, European Agency for Safety and Health at Work, Luxembourg, 2005.

7. A.C. Wong, K. Froud and Y. Hsieh, *World Journal of Otorhinolaryngology*, 2013, **3**, 3, 58.

8. G. Stevens, S. Flaxman, E. Brunskill, M. Mascarenhas, C.D. Mathers and M. Finucane, *European Journal of Public Health*, 2011, **23**, 1, 146.

9. *Tackling the Compensation Culture: Noise Induce Hearing Loss Claims*, Association of British Insurers, London, UK, 2015.

10. *Proceedings of the Listen Up! Symposium*, 2nd March, Manchester, UK, Health & Safety Laboratory, Buxton, UK, 2016.

11. *http://www.hsl.gov.uk/listenup*

12. *WHO Community Noise Guidelines*, World Health Organization, Geneva, Switzerland, 1999.

13. *The Costs and Benefits of the Noise at Work Regulations 1989*, HSE Contract Research Report 116/1993, Health and Safety Executive, Bootle, Merseyside, UK, 1993.

14. BS EN ISO 7029:2000 Acoustics – The statistical distribution of hearing thresholds as a function of age.

15. *Code of Practice Reducing the Exposure of Employed Persons to Noise*, Department of Employment, UK, 1972.

16. BS ISO 1999:2013 – Acoustics – Estimation of noise-inducting hearing loss.

17. EC Directive 2003/10/EC – Noise, European Commission, Brussels, Belgium, 2003.

18. EC Directive 2006/42/EC – Machinery, European Commission, Brussels, Belgium, 2006.

19. EC Directive 2000/14/EC – Noise – Equipment for use outdoors, European Commission, Brussels, Belgium, 2000.

20. *The Supply of Machinery (Safety) Regulations 2008*, Statutory Instruments No.1597, UK, 2008.

21. 'Noise Mapping in Paper Mills', Paper and Board Information Sheet No.2, Health and Safety Executive, Bootle, Merseyside, UK.

22. B. Ramazzini in *De Morbis Artificum Diatriba*' (Diseases of Workers), Modena, Italy, 1700.

23. *The Wilson Report on Noise*, Wilson Committee, March 1963.

24. *Noise at Work Regulations 1989*, Health and Safety Executive, Bootle, Merseyside, UK, 1989.

25. *Noise and the Worker*, UK Ministry of Labour, London, UK, 1963.

26. *Guide to the Exposure to the Exposure of Employed Person at Work*, Department of Employment Code of Practice, 1972.

27. *Health and Safety at Work Act*, Health and Safety Executive, Bootle, Merseyside, UK, 1974.

28. *100 Practical Applications of Noise Reduction Methods*, Health and Safety Executive, Bootle, Merseyside, UK, 1983.

29. *Sound Solution Case Studies*, Health and Safety Executive, Bootle, Merseyside, UK, 1995.

30. *Noise in the Plastics Processing Industry: A Practical Guide to Reducing Noise from Existing Plant and Machinery*, RAPRA, Shawbury, UK, 1985.

31. R.J. Peters in *Noise Control (A Pira Environmental Guide)*, Pira, Leatherhead, UK, 2000.

32. ISO 8253-1:2010 Acoustics – Audiometric test methods – Part 1: Pure-tone air and bone conduction audiometry.

5 Hearing Protection and the Use of Personal Hearing Protectors

5.1 Introduction

Professionals in occupational safety and health recommend a hierarchy of controls to protect the hearing of employees in the workplace:

- Elimination – physically remove the hearing hazard.

- Substitution – replace the hearing hazard.

- Engineering controls – isolate people from the hearing hazard.

- Administrative controls – change the way people work.

- Personal hearing protection – protect the worker with personal hearing protection.

5.1.1 Hierarchy of Hearing Protection Controls

Hearing protectors should reduce the risk of hearing damage to the wearer provided that they:

- Have been carefully selected by appropriately trained personnel to provide adequate noise attenuation against the noise to which the wearer is being exposed.

- Are in good working order, not worn or damaged, and well maintained.

- Wearer has been trained in the use of hearing protectors and that they have been fitted appropriately.

- Are worn for 100% of the time for which the wearer is exposed.

Even so, hearing protection should be a last resort when it is not possible to reduce the noise exposure of the wearer by other means, such as reduction of noise exposure levels or noise-exposure time, and that a programme of audiometric monitoring is in place as a check on the effectiveness of the use of hearing protection.

Therefore, hearing protection (sometimes called 'ear protection') should be seen by managers in industry as: a last resort for protecting people's hearing; a failure to reduce noise levels by other means; an interim method until a more acceptable method of controlling their exposure to noise is in place.

Nevertheless, the effective use of hearing protectors is likely to be the only practicable means available to protect the hearing of very large numbers of employees for several years to come. It is, therefore, very important that a well-managed programme is in place to ensure that they are always used effectively.

This chapter discusses the types of hearing protection available, their performance, issues relating to their selection and use (including difficulties and limitations) and the need for information, instruction and training for employees.

This chapter has used material contained within part 5 of the Health and Safety Executive (HSE) Guidance on the Regulations, Controlling Noise at Work L108 under *the Control of Noise at Work Regulations 2005*.

5.2 Types of Hearing Protectors

Earmuffs are fitted over the ears to block the path of airborne sound to the ear canal.

They consist of four main parts:

- Moulded rigid plastic cups which fit over the ears and which provide the basic sound insulation.

- Sound-absorbing foam plastic lining fitted inside each cup to reduce reverberation.

- Flexible plastic cushions designed to enable the cups to fit closely over the ears and establish a close seal to the wearer's head to minimise leakage of sound around the cups to the ear. The cushions are made out of thin flexible plastic film, and most types are filled with foam plastic, but there are some which are filled with oil.

- An adjustable elastic, flexible headband which holds the cups in place and helps the cushions to make a good seal to the head.

Although all earmuffs conform to this same basic pattern, there is a variety of different cup sizes and weights. Muffs are available which fit onto 'hard hats'.

Earplugs are fitted into the ear canal. There are many different types, made from soft rubber, foam plastic or a special type of mineral fibre called 'glass down' – note that ordinary cotton wool is not effective. In each case, they are designed to completely fill the ear canal and, therefore, minimise the sound transmitted to the ear drum. It is, therefore, important that the types selected are a good fit and are inserted appropriately according to manufacturer instructions.

Headband-mounted ear plugs (sometimes called 'canal caps') consist of a pair of plastic or rubber plugs connected by a light flexible plastic headband. They are designed to be quick, easy to fit, and remove for personnel who frequently move in and out of noisy areas, such as managers and supervisors.

5.2.1 Specialist Types

Specially developed amplitude-sensitive earmuffs are available which allow low-amplitude sound, for example, from normal conversation, but prevent transmission of high intensities. These are useful if there are sudden, unpredictable bursts of high levels of noise.

Other types of plugs are available which are moulded individually and fitted with an acoustic filter designed to ensure that the device produces more or less the same attenuation at all frequencies. This type of plug is used, for example, by musicians when it is necessary that the frequency spectrum of the sound being listened to, and so its character, is preserved.

Earmuffs fitted with communication headsets are also available so that employees can enjoy protection from high levels of machine noise, listen to 'piped music', and receive messages from the workplace communications system.

Earmuffs fitted with active sound attenuation are available. That is, they contain a miniature microphone and loudspeaker which detects the sound waveform and provides a cancelling 'anti-phase' signal.

5.3 Noise-attenuating Performance of Hearing Protectors

Suppliers of hearing protection with the Conformité Européenne (CE) mark are required to satisfy the relevant part of BS EN 352, which sets out basic safety requirements for hearing (e.g., features such as size, weight and durability) for:

- Earmuffs (BS EN 352-1:2002);
- Earplugs (BS EN 352-2:2002); and
- Helmet-mounted earmuffs (BS EN 352-3:2002).

Hearing protection which complies with BS EN 352 must be supplied with performance information derived from a standard test defined in BS EN 13819-2:2002 (which in turn draws on a method in BS EN 4869-1:1993). The information required is:

- Mean and standard deviation (SD) attenuation values at each octave-band centre frequency from 125 to 8 kHz (63 Hz is optional);

- Assumed protection values (APV) at each centre-frequency (based on the mean minus one SD);

- High (H), medium (M) and low (L) frequency values in accordance with BS EN ISO 4869-2:1995; and

- Single number rating (SNR) value in accordance with BS EN ISO 4869-2:1995.

The high, medium, and low method (HML) for rating hearing protectors and SNR values are derived from the mean and SD attenuation values, and are explained below.

The first requirement of a hearing protector is that it should provide sufficient noise attenuation to reduce the noise at the wearer's ear to a level which will prevent the risk of noise-induced hearing loss (NIHL). This depends on the levels and frequency spectrum of the noise.

The sound pressure levels at the ear when hearing protection is worn may be estimated using several methods. The principal three methods for passive hearing protectors are defined in BS EN ISO 4869-2:1995: octave-band method, HML method, and SNR method.

The first method requires data about the noise level to which the unprotected ear is exposed in octave bands. The two other methods do not require octave-band data; the HML method requires both the overall A-weighted sound pressure levels (L_A) and C-weighted sound pressure levels (L_C); and the SNR method requires only the C-weighted value.

All methods will give similar predictions of sound levels at the ear for general industrial and occupational noise sources. The HML and SNR methods become less accurate when compared with the octave-band method if the noise is dominated by noise at single frequencies, particularly if these are at low frequencies.

5.3.1 Octave-Band Method

The performance of ear protectors is specified as a mean attenuation measured in each of seven octave bands (from 125 Hz to 8 kHz) according to a test method specified in BS EN 24869-1:1993 and ISO 4869-1:1990, according to which the hearing thresholds of test subjects is measured with and without hearing protectors, with the attenuation being the difference between the two measurements. The attenuation obtained from individual test subjects will be different because of differences in the quality of fit, and the test states that the SD from the mean should also be specified. If the mean value of the attenuation were to be used in calculations of noise exposure, then half the population of wearers would receive a lower value of protection than predicted. Therefore, the method given in BS EN 24869-1:1993 requires that an APV of attenuation of the mean value minus the SD be used in such calculations of noise exposure.

Assuming that the spread of attenuation values follows Gaussian statistics, 68% of the wearing population will receive the assumed level of protection. If a value of mean minus two SD were to be used, then 95% of the wearing population will receive the calculated level of protection.

To calculate the amount of protection afforded in any particular application, it will be necessary to measure the level of noise at the wearer's ear, in octave bands. The assumed protection provided by the protector in each band is then subtracted to obtain the band level at the 'protected' ear. These levels are then A-weighted, and combined, to give the level at the protected ear, in dBA.

5.3.2 HML Method

Manufacturers provide three different performance figures to be used depending on the frequency spectrum of the noise. The noise level is measured using both the A- and C-frequency weighting scales, and the difference between the two readings $(L_A - L_C)$ is indicative of the type of spectrum. The H, M and L values are the predicted noise-level reduction (PNR) of the hearing protector for noises with values of $(L_A - L_C)$ equal to –2, 2 and 10 dB, respectively. If the difference $(L_A - L_C)$ is different from one of these three values, the value of PNR is obtained by interpolation using **Equation 5.1** or **5.2**:

If $(L_A - L_C) > 2$ dB,

$$PNR = M - [\{(M - L)/8\}(L_A - L_C - 2)\tag{5.1}$$

Otherwise:

$$PNR = M - [\{(H - M)/4\}(L_A - L_C - 2)$$ (5.2)

The effective A-weighted sound pressure level at the protected ear (L'_A), is then given by **Equation 5.3**:

$$L'_A = L_A - PNR$$ (5.3)

5.3.3 SNR Method

Manufacturers provide an SNR value for each type of hearing protector. The method requires measurement of the L_C.

The L'_A is given by subtracting the SNR value for the protector from the L_C, **Equation 5.4**:

$$L'_A = L_C - SNR$$ (5.4)

Example

The sound pressure level close to a machine in a printing works is 96 dBA and the octave-band spectrum of the noise is given below. Using the octave-band method, the HML method and SNR method are used to calculate the assumed protected level at the ear of an employee working near to the machine who is wearing ear protectors having the octave-band performance given in the **Table 5.1**, and with H = 25, M = 17, L = 15 and SNR = 21 dB.

Table 5.1 Calculation using the octave-band method							
1	2	3	4	5	6	7	8
Octave band (Hz)	Octave band SPL (dB)	Mean attenuation (dB)	Standard deviation (dB)	Assumed protection (dB)	Octave band APL (dB)	Octave band A-weighted (dB)	A-weighted octave band APL (dB)
63	79	20	4	16	63	−26	37
125	82	19	3	16	66	−16	50
250	87	18	5	13	74	−9	65
500	82	18	6	12	70	−3	67
1,000	91	21	4	17	74	0	74
2,000	92	31	4	27	65	+1	66
4,000	87	37	4	33	55	+1	55
8,000	86	38	6	32	54	−1	53

- Octave-band method

  The assumed protection of column 5 is given by subtracting the SD of column 4 from the mean attenuation of column 3. The A-weighted protected octave-band level of column 8 is obtained by subtracting the APV of column 5 from the band level of column 2, to give column 6, and then adding the A-weighting value of column 7. The overall protected level is obtained by combining the levels in column 8 using the rules of dB arithmetic (**Chapter 1**), to give an assumed protected level of 76 dBA.

  Thus, for this particular application, the overall protection provided in reducing the level at the ear is from 96 to 76 dBA.

- HML method is given in **Equations 5.5–5.7** using:

  L_A = 96 dB and L_C = 96 dB

  Therefore:

$$L_A - L_C = 0 \tag{5.5}$$

  and:

$$PNR = M - [\{(H - M)/4\}(L_A - L_C - 2)] = 17 - [(25 - 17)/4\}(0 - 2)] =$$

$$17 - (8/4)(-2) = 21 \text{ dB} \tag{5.6}$$

and:

$$L'_A = L_A - PNR = 96 - 21 = 75 \ dB \qquad (5.7)$$

- SNR method becomes **Equation 5.8**:

$$L'_A = L_C - SNR = 96 - 21 = 75 \ dB \qquad (5.8)$$

5.3.4 Optimum Range for the Sound Pressure Level at the Protected Ear

Ideally, the hearing protection should reduce the sound pressure level at the ear to well below 85 dBA and preferably in the range 75 to 80 dBA.

However, it is the exposure level (which depends upon the duration as well as the sound level) which is important, and so higher levels may be acceptable for shorter periods. The *Control of Noise at Work Regulations 2005* requires that the protectors should provide sufficient attenuation to reduce exposure levels to below the upper-exposure action value of 85 dBA, although the lower-exposure action value level would be a safer objective.

5.4 Inspection, Care, Maintenance and Replacement

After prolonged use: plastic plugs may become inflexible; the seals of earmuffs may become worn or hard and, therefore, ineffective; headbands become may lose their elasticity; the cups may become damaged. In all cases, the degree of protection will be drastically reduced until the protectors are repaired or replaced. Training must, therefore, cover inspection of protectors to alert wearers to these defects.

5.5 Accounting for 'Real World' Protection

Research has shown that in real use the protection provided can be less than predicted by manufacturer's data. To give a realistic estimate, allowing for the imperfect fitting and condition of hearing protectors in the working environment, it is recommended that a real-world factor of 4 dB is applied. That is, that 4 dB is added to the estimated sound levels at the protected ear calculated by any of the above three methods.

5.6 Dual Protection

Sometimes employees may be exposed to very-high noise levels, and the above calculations may show that earplugs or earmuffs will not reduce the noise level at the ear to a low enough level. This problem is likely if the daily noise exposure is more than 110 dB or the peak sound pressure level is more than 150 dB, especially if there is substantial noise at frequencies less than 500 Hz. If muffs and plugs are worn together, then this may provide an increased amount of protection, but only by a few extra dB, and it is not easy to predict the combined performance. The amount of protection will depend on the particular combination of earmuffs and earplugs. In general, the most useful combination is a high-performance earplug with a moderate-performance earmuff (a high-performance earmuff adds a little extra protection but is likely to be less comfortable).

If dual protection is used, test data should be obtained for the particular combination of earplug and earmuff (and helmet, if used). In practice, the increase in attenuation to be expected from dual protection will be less than 6 dB over that of the better of the individual protectors.

Under these circumstances, it is strongly recommended that alternative working procedures are found, and that exposure times at these levels are reduced to an absolute minimum and are strictly monitored, even with ear protectors being worn.

5.7 Overprotection

Hearing protectors that reduce the level at the ear to less than 70 dB should be avoided because such over-protection may cause difficulties with communication and hearing warning signals. Users may become isolated from their environment, leading to safety risks, and generally may have a tendency to remove the hearing protection and, therefore, risk damage to their hearing.

5.8 Selection of Hearing Protectors

5.8.1 Adequate Noise-Attenuating Performance

For very high-noise levels (above 100 dBA), the lower-exposure action value would be reached within a few minutes. In such cases, it should not be assumed that wearing a particular type of hearing protector will provide adequate noise attenuation. It will be necessary to use one of the three methods described above to check that the protectors being used provide sufficient protection.

At lower levels of exposure, for example, 92 or 93 dBA, any reputable earplug or earmuff with the CE mark indicating compliance with the relevant part of BS EN 352 should provide ample protection if in good condition and fitted appropriately. In these situations, emphasis should be placed on providing a selection of types for the wearer to choose from to reduce discomfort and inconvenience to a minimum and, therefore, to maximise the chances that the protectors will be worn for 100% of the time for which they should be worn.

5.8.2 Compatibility, Comfort, Convenience and Hygiene

Earmuffs are easier to fit than earplugs, and they can be worn by almost everyone. Some people may have a medical condition which prevents them from wearing earplugs, for example, they may cause irritation of the ear canal. Earmuffs generally, but not invariably, provide higher levels of attenuation than earplugs (depending upon the particular types) and this attenuation is less variable because less skill is required in fitting them appropriately. Earmuffs can be inconvenient, uncomfortable, hot and heavy, particularly if they have to be worn for several hours in warm and restricted spaces. Earplugs will be lighter and more comfortable in such circumstances.

Good hygiene is essential if earplugs are to be used, and hands should always be washed before inserting or removing the earplugs into the ear canal. Earmuffs are, therefore, preferable in dirty areas where adequate washing facilities are not available, or if they have to be frequently taken on and off because the wearer is moving in and out of hearing-protection zones. People who wear glasses or earrings, or who have long hair may receive a reduced amount of protection from earmuffs if the quality of fit of the seals of the muffs to the head is reduced. Also, there can be problems if other safety equipment has to be worn, such as safety glasses, face masks or other respiratory equipment, although special types of earmuffs are available to deal with some of these situations (e.g., earmuffs attached to safety helmets).

5.9 Difficulties and Limitations of Hearing Protectors

The difficulties are that they may not be worn and, even if they are worn, they may not be giving the estimated degree of protection because they have not been fitted correctly or because they are defective in some way.

Many employees understandably do not like having to use ear protectors, particularly if they have to be worn for several hours each day. Apart from the discomfort and inconvenience, several other reasons for concern are often given, including that they impair the wearer's ability to detect changes in the noise emitted by the machines they are operating, or to hear alarm signals, communications from fellow workers, or the public address system. In fact, the protectors reduce both the level of the background noise as well as of the wanted signal. It has been shown that for wearers with good hearing, their detection of these signals is undiminished or even increased, but there are problems for wearers who have a hearing impairment. Sometimes those already suffering from NIHL put forward objections to wearing ear protectors on the basis that they are not needed because they have 'got used to the noise', or alternatively that 'it is too late' because they already have a hearing impairment. The guiding principle in such cases should be to safeguard all the more carefully the hearing that remains. It is important to counter and overcome all such concerns with a programme of training that explains the need to wear ear protectors and the consequences of not doing so.

The importance of the need for the protection to be worn for 100% of the time exposed to high levels of noise cannot be overestimated. No matter how good the assumed protection provided, and this can be in the region of 20 of 30 dB, it will be reduced to only 3 dB if they are only worn for 50% of the time and, even if they are worn for 90% of the time, the degree of protection will only be 10 dB. This is particularly important if the noise levels are very high because the exposure level increases rapidly even in a short time. At a level of 102 dBA, for example, a personal daily noise-exposure level of 85 dBA (the upper-exposure action level in the *Control of Noise at Work Regulations 2005*) is reached after 30 min, and this falls to 15 min at 105 dBA and 7.5 min at 108 dBA. In these circumstances, it is obvious that failure to wear protectors even for a few minutes will be harmful. It is important to explain and emphasise this degree of urgency in training sessions.

The attenuation figures provided by manufacturers are obtained from tests under standardised measurement conditions. They represent the maximum attenuation likely to be met under laboratory conditions when protectors are new and always fitted appropriately. Much lower levels of protection are obtained if the protectors are not fitted appropriately. This is particularly important with earplugs, which must always be inserted exactly according to manufacturer's instructions. Training is needed to demonstrate the appropriate procedure and to ensure that each employee can fit the earplugs appropriately, and even then they may become loose over time. Even with earmuffs, which are much easier to fit appropriately, it is necessary to sweep back the hair, ensure that the cups are completely surrounding the ears, and adjust headband tension to ensure a good seal with the side of the head. It has been shown that a few minutes spent demonstrating and explaining these points leads to higher levels of attenuation.

5.9.1 Costs of Ear Protectors

Although the cost of each earmuff or earplug is relatively small, the cost of their continued provision for large numbers of employees over several years, together with the cost of managing their use, including the time spent on training and monitoring and record-keeping, can be substantial and comparable with the costs of investing in noise-control measures.

5.10 Two Health and Safety Executive Reports on Hearing Protection

In 2002 the HSE commissioned a study into *Behavioural Studies of People's Attitudes to Wearing Hearing Protection and How These Might be Changed* (HSE Research Report 028).

The overall aim of the project was to investigate how industry has attempted to motivate employees to wear hearing protectors and to test the effectiveness of these methods.

Information was gathered *via* questionnaires sent to employees and employers. A total of 280 questionnaires were collected from 19 companies.

The companies surveyed were from a range of industries and covered large, medium, small and very small employers. There was a range of different management approaches to noise control and generally the larger companies had effective or partly effective hearing-protection programmes in place. The smaller companies had very limited noise-control procedures and relied heavily on personal protective equipment.

The interventions carried out in Phase 2 were designed to address the specific needs of four separate workplaces, previously surveyed in Phase 1. These interventions included basic noise-awareness training, provision of alternative types of hearing protection, and coaching of management in basic feedback and communication techniques for encouraging workers to modify their behaviour. The interventions were designed to be participative and involving. The results of the interventions were assessed by comparing the observed behaviour with that previously noted in Phase 1 and also by administering a post-intervention questionnaire during a follow-up survey 8 weeks after the workplace interventions had been carried out.

In all cases, the interventions showed positive results, with increased hazard awareness among the workers and increased use of the hearing protection. The most basic types of intervention, such as noise-awareness training and provision of the most suitable hearing protection for the job, showed the greatest improvement. The feedback from the behavioural safety training that was carried out was encouraging but the improvements were subtle and difficult to detect.

The main conclusions and recommendations related to the importance of adequate selection of the type of hearing protection, to management commitment to the hearing-protection programme, and about the quality of information, instruction and training given to employees.

Face-to-face training was considered to be the most effective training method, particularly regarding the correct fitting of ear protectors, with employee participation and time for questions. It is no use issuing hearing protection and then expecting the workers to use it correctly. In particular, workers need to be shown how to insert earplugs correctly so that they get the maximum benefit. This should be demonstrated by a trained and experienced user.

5.10.1 Real-World Use and Performance

A second report, a HSE Research Report RR720 on *Real World Use and Performance of Hearing Protection*, was issued in 2009.

This report considered the effectiveness of hearing protectors in everyday work situations. The study was undertaken in two parts. The first consisted of interviews with employers to discuss management of noise and hearing-protector use, and on-site observation of hearing-protector use. The purpose of these visits was to see: how well hearing protection was used; the training provided; the use of other personal protective equipment and equipment that may limit attenuation; behavioural factors affecting use taking into account the noise exposure of employees and the environment in which the hearing protection is worn.

The second part of the study was objective laboratory measurements of loss of insertion of the hearing protector. The purpose of these measurements was to quantify the reduction in protection due to poor fitting or maintenance for a range of hearing protectors.

Some of the finding related to poorly fitted or improperly used hearing protectors:

- Compressible foam earplugs are generally poorly fitted because users are unaware of how these should be compressed before fitting, or unaware of the importance of correct compression. An incorrectly compressed earplug may give virtually no attenuation.

- There is a need for clothing compatible with correct fitting of earmuffs for outdoor workers – earmuffs worn over conventional hats and hoods can provide only limited attenuation and are likely to leave the wearer unprotected.

- Banded ear canal caps must be partially inserted into the ear canal as shown by the manufacturer's fitting instructions. Caps held only against the canal entrance by the band may give no protection.

- Laboratory tests show that the tension of earmuff headbands reduces with use and stretching, and that a reduction of 6 dB from the manufacturer's SNR value could be typical of earmuffs with only moderate use.

5.11 The Need for Health Surveillance – Routine Hearing Tests

Therefore, it can be seen that if a company's hearing-protection policy relies heavily on the use of ear protectors, particularly if there are high levels of noise, then for all of the reasons described above there will be some risk that employees may not be receiving the amount of protection predicted on the basis of assumed protection calculations. Hence, health surveillance in the form of regular audiometric monitoring is used as a check on the effectiveness of the hearing-protection policy, as required in the *Control of Noise at Work Regulations 2005*.

More detailed information about hearing protectors is available in part 5 of the HSE Guidance on the Regulations, Controlling Noise at Work L108 under *the Control of Noise at Work Regulations 2005*.

5.12 Information, Instruction and Training

Training is needed to cover the aspects indicated below:

- Why protectors should be worn (i.e., training should include an explanation of NIHL and its consequences). There is a risk of hearing loss and its possible effects on the individual, and also any legal and disciplinary consequences. The persuasive arguments relating to the personal and social consequences of loss of hearing should be presented, and not just the more coercive approach emphasising duties relating to the law and/or company policy.

- Where and when hearing protectors should be worn (i.e., in which areas of the workplace and when certain machines or processes are in operation). This information should also be clearly marked by signs at the entrance to all hearing-protection zones. Information about noise levels and noise exposure levels should be provided or made available.

- How to obtain ear protectors (i.e., from where and from whom in the organization).

- Introduction to a range of different types, with as far as possible, the opportunity to exercise personal preference regarding convenience and comfort.

- Careful instruction into correct fitting procedures for both earmuffs and earplugs.

- Instruction about care of protectors (e.g., washing and cleaning) and how to inspect to ensure that they are still effective, and when to replace.

- Importance of wearing for 100% of the time in hearing-protection zones.

- Duties of employees under the *Control of Noise at Work Regulations 2005* and any specific conditions of employment relating to the wearing of ear protectors.

- Where to obtain further information and whom to see about any future queries and concerns.

- Training should be given to managers as well as shop-floor employees. It is particularly important, for motivational reasons, that senior managers in the company are seen to be actively involved in the hearing-conservation policy, and that they always wear ear protectors when in a hearing-protection zone, even if they are present for only a short period of time. Provision for the training of new employees should be made. Monitoring and supervision should be used to identify when further updating training, and reinforcement training is needed either on a collective or an individual basis.

- The provision of information, instruction and training is not only an important duty of employers under the *Control of Noise at Work Regulations 2005*, but also an important defense against possible future civil law claims from employees and former employees for loss of hearing caused by noise exposure at work. It is, therefore, essential that complete records are kept of all training activities.

5.13 Hearing Protectors – Summary and Conclusions

The main points covered in this chapter together with conclusions are summarised below:

1. The use of hearing protectors is at the bottom of the recommended hierarchy of controls (i.e., as a last resort).

2. This chapter discusses the types of hearing protection available, their performance, issues relating to their selection and use (including difficulties and limitations) and the need to provide information, instruction and training for employees.

3. The main types are earmuffs, earplugs, and specialist types compatible with the use of other types of personal protective equipment, such as safety goggles and safety helmets.

4. There are three main methods for specifying the sound-attenuation performance of hearing protectors: the octave-band method, HML method and SNR method. The three methods are explained and illustrated with a worked example.

5. To ensure that hearing protectors are always effective, it is important that they are cared for appropriately, regularly inspected, maintained and replaced when necessary.

6. An allowance should be made of a reduction of 4 dB when estimating the sound-attenuating performance of hearing protectors, to allow for real-world protection.

7. Dual protection – if dual protection is used, test data should be obtained, if possible, for the particular combination of earplug and earmuff (and helmet, if used). In practice, the increase in attenuation to be expected from dual protection will be less than 6 dB over that of the better of the individual protectors.

   Under these circumstances it is strongly recommended that alternative working procedures are found, and that exposure times at these levels are reduced to an absolute minimum, and are strictly monitored, even with ear protectors being worn.

8. Overprotection – to prevent the difficulties associated with feelings of isolation, with communication and warning signals, hearing protectors that reduce the level at the ear to less than 70 dB should be avoided.

9. Hearing protectors should be selected to be suitable and effective taking into account the need to provide sufficient noise attenuation for the wearer and factors such as comfort, convenience, hygiene considerations and compatibility with

the task being performed, and with other personal protective equipment such as goggles or safety hats.

10. Difficulties and limitations of hearing protectors – the main difficulties are that employees may not use their hearing protectors for 100% of the duration of their noise exposure, and that the protectors may not be providing as much attenuation as estimated because of incorrect fitting and because protectors may be worn or damaged.

11. Two investigations have reported the effectiveness of hearing protection. They have confirmed and highlighted many of the shortcomings and limitations in the use of hearing protectors described above.

12. The need for health surveillance – routine hearing tests. For all of the reasons described above, there will be some risk that employees may not be receiving the amount of protection predicted on the basis of assumed protection calculations.

    Hence, health surveillance, in the form of regular audiometric monitoring, is used as a check on the effectiveness of the hearing-protection policy, as required in the *Control of Noise at Work Regulations 2005*.

13. Information, instruction and training – employees should provide information, instruction and training for employees so that they understand why, how, where and when hearing protection should be worn.

14. The costs of providing hearing protectors – although the cost of each earmuff or earplug is relatively small, the cost of their continued provision for large numbers of employees over several years, together with the cost of managing their use (including the time spent on training and monitoring and record-keeping) can be substantial and comparable with the costs of investing in noise-control measures.

15. More detailed information about hearing protectors is available in part 5 of the HSE Guidance on the Regulations, Controlling Noise at Work L108 under *the Control of Noise at Work 2005 Regulations*.

5.14 Conclusions

The use of earmuffs and earplugs may be a necessary last resort to protect employees' hearing, but their use as an interim measure must not detract from efforts to reduce noise exposure from other, more acceptable and reliable means. The widespread use of ear protection by employees needs to be managed properly, is not necessarily a cheap and simple option, and has limitations and difficulties. It is essential that employees are fully informed, trained and supervised in relation to the use of ear protection.

There are legal requirements under the *Control of Noise at Work Regulations 2005* relating to the provision of ear protectors.

Bibliography

1. Controlling Noise at Work L108 under *the Control of Noise at Work 2005 Regulations*, Guidance on the Regulations, Health and Safety Executive, Bootle, Merseyside, UK, 2005.

2. BS EN 352 – Hearing protectors.

3. BS EN 352-1:2002 – Hearing protectors – Safety requirements and testing – Ear-muffs.

4. BS EN 352-2:2002 – Hearing protectors – Safety requirements and testing – Ear-plugs.

5. BS EN 352-3:2002 – Hearing protectors – Safety requirements and testing – Ear-muffs attached to an industrial safety helmet.

6. BS EN 13819-2:2002 – Hearing protectors – Testing – Acoustic test methods.

7. BS EN 24869-1:1993, ISO 4869-1:1990 Acoustics – Hearing protectors – Sound attenuation of hearing protectors – Subjective method of measurement.

8. BS EN ISO 4869-2:1995 – Acoustics – Hearing protectors – Estimation of effective A-weighted sound pressure levels when hearing protectors are worn.

9. G.W. Hughson, R.E. Mulholland and H.A. Cowie in *Behavioural Studies of People's Attitudes to Wearing Hearing Protection and How These Might Be Changed*, HSE Research Report 028, Health and Safety Executive, Bootle, Merseyside, 2002.

10. L. Brueck in *Real World Use and Performance of Hearing Protection*, HSE Research Report RR720, Health and Safety Executive, Bootle, Merseyside, 2009.

11. *Control of Noise at Work Regulations 2005.*

6 Noise in the Environment

6.1 Introduction

In addition to causing problems for people in the workplace, noise generated in the workplace can also cause disturbance to people nearby.

These include people at work and at leisure outside in gardens, parks, public spaces as well as those living and working within nearby buildings. Particularly noise-sensitive properties are residential dwellings, schools and hospitals. Noise may originate from: machines and processes within workplace buildings (in which case the fabric of the building envelope will provide some attenuation of noise transmitted to the environment); external sources such as generators, pumps, motors, and heating, ventilation and air conditioning (HVAC) equipment: transport of material to and from the workplace, and associated loading and unloading of vehicles, and use of fork lift trucks; sound from public address systems such as staff announcements, alarms, and end-of-shift signals.

This chapter will briefly review target noise levels which should be achieved within the environment not to cause disturbance to people living and working nearby.

It will outline strategies from minimising noise emission and suggest relevant standards and codes of practice relating to industrial noise in the environment, and noise disturbance.

6.2 Regulations, Standards and Codes

6.2.1 Nuisance and Planning

Noise from industrial sites affecting nearby communities will be subject to control by nuisance and planning legislation. Noise nuisance is dealt with by the Environmental Health Department under the provisions of the Environmental Protection Act, 1990. Under these provisions, any noise considered to be 'nuisance' becomes a statutory nuisance and must be abated. The emission of noise from existing industrial sites may be controlled by planning conditions, enforced by local planning authorities, and new industrial sites (or changes to noise sources at existing sites) will require application to be made for planning permission, including a report on the noise impact of the proposed developments.

Noise-control measures required for new planning permission are likely to be more demanding than those for control of noise from an existing site for solving a nuisance dispute.

6.2.2 Integrated Pollution Prevention and Control Regulations

In addition to nuisance and planning legislation, certain industrial processes are required to meet permitted noise levels specified under the Integrated Pollution Prevention and Control (IPPC) regulations.

Industrial units which operate under the IPPC system will need to apply for and obtain a permit for their noise emissions. To gain a permit, operators will have to show that they have systematically developed proposals to apply the best available techniques and meet certain other requirements by taking account of relevant local factors.

Once a permit has been issued, other parts of IPPC come into play. These include compliancemonitoring, periodic permit reviews, variation of permit conditions and transfers of permits between operators.

Depending on the size and complexity of the operations, these will be enforced either by the Environment Agency or the local authority Environmental Health Department.

6.3 Criteria and Noise Targets for Environmental Noise Emissions

It is usual to specify noise limits outside noise-affected buildings rather than inside them because it is much easier to verify compliance by carrying out noise measurements outside, and indeed it would not usually be possible to obtain access for internal measurements to be taken. The noise limit, which relates to human responses to noise, applies inside, so it is necessary to add a factor for the sound insulation provided by the building fabric. In the summer, when windows are open, a rule of thumb for this is 10–15 dBA, although sometimes a more accurate estimate may be required.

Noise limits will usually be specified in terms of an equivalent (or average) sound (or noise) level in decibels (dB) over a period of time (LAeq,T) value, supplemented by the maximum noise level to deal with the effect of a short-duration burst of high-level noise [maximum sound level(s) (LAmax) values] especially at night.

6.4 Noise-Impact Assessment Methods

6.4.1 BS 4142:2014

BS 4142:2014 is the most commonly used method for noise-impact assessments of industrial noise in the UK. It provides a method for assessing the impact of industrial sound on people in nearby buildings. The method is based on the difference between the existing ambient noise level at the noise-sensitive premises and the sound level arising from the introduction of the new sound source, called the 'specific sound level'. The idea is that a small difference is considered to represent a small impact and may be acceptable, whereas a much larger difference will represent a much larger impact and will be likely to give rise to an adverse impact, and be unacceptable. The standard is intended for use in informing planning decisions involving industrial and commercial sites.

BS 4142:2014 refers throughout to 'sound' rather than 'noise', but recognising that some sounds are unwanted, and so may also be regarded as and referred to as noise.

The specific sound level is measured (or predicted) as an LAeq,T value over a specified time interval (1 h for day, and 15 min for night-time). The ambient sound level is measured as the L_{A90} value of the ambient sound, called the 'background sound level'.

The specific sound level should be measured outside the property (or properties) likely to be affected by the specific sound and at more than 3.5 m from their façades to avoid sound reflection from the façades affecting the measurement.

The measured values of the specific sound levels may be subject to three possible corrections: (i) for the influence of the level of the ambient sound; (ii) for the effect of the duration of the measurement sample [e.g., in cases where the specific sound occurs in short bursts which are less than the specified time interval (e.g., less than 1 h in the daytime)]; and (iii) for the character of the sound, to take into account the effect of features in the sound which may increase the annoyance that it might cause. The value of the specific sound when subjected, as appropriate to none, some, or all of these three types of correction, becomes the rating level of the specific sound. It is this value which is then compared with the measured background sound level to determine the assessment (i.e., the degree of adverse impact).

To determine whether the first of these corrections (the effect of ambient sound level on the measurement of the specific sound level) is required, it is necessary to measure the LAeq,T value of the ambient sound (which is called the 'residual sound level') when the specific sound level is not present. If the residual sound level is less than 10 dB below the specific sound level, then the residual sound (which was present during the measurement of the specific sound level) will have had a significant influence on the measured specific sound level, and a correction must be made, as described in the standard.

6.4.1.1 Character Corrections

There are three possible acoustic characteristics which may be taken into account in making the assessment: tonality, impulsivity, and any other features likely to attract attention (e.g., intermittency). The presence and degree of importance of these features should preferably be agreed on the basis of subjective judgements but, if necessary, objective measurements are provided to determine tonality or impulsivity.

6.4.1.2 Assessment Procedure

The significance of the sound of an industrial and/or commercial nature depends upon both the margin by which the rating level of the specific sound source exceeds the background sound level and the context in which the sound occurs:

- Typically, the greater this difference, the greater the magnitude of the impact.

- A difference of around +10 dB or more is likely to be an indication of a significant adverse impact, depending on the context.

- A difference of around +5 dB is likely to be an indication of an adverse impact, depending on the context.

- The lower the rating level is relative to the measured background sound level, the less likely it is that the specific sound source will have an adverse impact or a significant adverse impact. Where the rating level does not exceed the background sound level, this is an indication of the specific sound source having a low impact, depending on the context.

Adverse impacts include, but are not limited to, annoyance and sleep disturbance. Not all adverse impacts will lead to complaints and not every complaint is proof of an adverse impact.

If the initial estimate of the impact needs to be modified due to the context, all pertinent factors should be taken into consideration, including the: absolute level of sound; character and level of the residual sound compared with the character and level of the specific sound; sensitivity of the receptor and whether dwellings or other premises used for residential purposes will already incorporate design measures that secure good internal and/or outdoor acoustic conditions.

BS 4142:2014 gives full details of appropriate sound-measurement and impact-assessment procedures, including a range of examples of how it should be used; it should be consulted for further information. It is recommended that it should be used only by those with experience and training in the measurement and assessment of sound levels.

6.4.2 BS 7445/ISO 1996

BS 7445/ISO 1996 is another noise impact-assessment procedure. It is broadly similar to BS 4142:2014 in as much as both are based on the extent that a specific noise, measured as an LAeq,T value, exceeds existing ambient noise levels, but is much wider in scope, covering all types of environmental noise. BS 7445/ISO 1996 gives general guidance on noise measurement and assessment procedures and on the setting of noise limits but, unlike BS 4142:2014, does not suggest noise limits.

The standard is in three parts:

- Part 1: Guide to quantities and procedures;

- Part 2: Guide to the acquisition of data pertinent to land use; and

- Part 3 Guide to application to noise limits.

6.5 Criteria for Fixed Noise Levels

In addition to impact-assessment procedures based on the degree to which specific sound levels exceed existing ambient sound levels, there are also recommended sound or noise levels based on research into human responses to sound and noise.

The most relevant to assessment of industrial and commercial noise are:

- The World Health Organization (WHO) guidelines published in 1989. These recommend that noise levels outside buildings should not exceed 55 dBA in the daytime or 45 dBA at night. More recent WHO guidelines (*WHO Night Noise Guidelines for Europe 2009*) recommend working towards the lower limit of 40 dBA outside at night-time.

- BS 8233:2014 advises (in Table 4) that, in general, for steady external noise sources, it is desirable that the internal ambient noise level in living/dining rooms does not exceed the guideline values of 35 dB LAeq,T in the daytime [07:00 to 23:00 (16 h)] or 30 dB LAeq,T in bedrooms at night [23:00 to 07:00 (8 h)]. Guidance given in BS 8233:2014 is informed and based upon the WHO guidelines.

In summary, there are a range of target levels which are used, but typically 40 to 50 dBA outside residential properties in the daytime and 30 to 35 dBA at night-time are acceptable.

Where these guidelines relate to indoor noise levels it will be necessary to estimate noise levels external to the building using information about the sound insulation of the building. BS 8233:2014 gives guidance on how to do this, and also includes a worked example (page 65) for calculation of external noise intrusion.

The Institute of Environmental Management and Assessment has published a general review of the various different methods available for the assessment of environmental noise: *Guidelines for Environmental Noise Impact Assessment*, version 1.2, November 2014.

6.6 Strategies for Minimising Noise Emissions and Disturbance to Neighbours

Disturbance is most likely to occur when noisy equipment is frequently in operation that is within the line of sight between the source and receiver at the nearest noise-sensitive properties at the most sensitive periods.

Therefore, much can be done by fairly obvious common-sense measures such as:

- Locating noisy equipment well away from, and out of the line of sight of, noise-sensitive properties, for example by making use of shielding by existing buildings where possible, or by erecting purpose-built noise barriers. Avoid using noisy equipment at particularly sensitive times, such as evening, night-time, early mornings and weekends and public holidays. Switch off equipment when not needed or reduce speed (of ventilation fans, for example) where possible at times of reduced output; and

• For sources inside buildings, keep windows and doors closed, in good repair and well-sealed. Keep all noise-producing equipment and machines and all noise -control devices (such as enclosures, barriers, attenuator and vibration isolators) in good repair and well maintained.

6.6.1 Good Relations with Neighbours

Good relations and good communications between noise-emitting premises and their noise-sensitive neighbours can help reduce levels of dissatisfaction and complaint. People may be more tolerant and understanding if they know what is happening and why certain noisy processes are important and necessary. This can include giving advance warning when new noisy activities are due to occur and providing apologies and explanations when unexpected problems lead to unexpected bursts of noise. If people are forewarned about noise events such as high-pressure jet releases, or testing of emergency generators, for example, they can prepare themselves and make arrangements to minimise their disturbance.

An effective procedure for noise complaints should be in place, with the appropriate staff trained to deal with complaints about noise effectively and sympathetically. Each complaint should be investigated thoroughly and promptly, and with complainants being kept updated of progress. Records of complaints, and how they were resolved, should be maintained.

Planning to meet environmental noise targets involves the following steps:

• Predict noise level external to the nearest noise-sensitive property;

• Compare predicted noise level with noise target level or noise criterion;

• If necessary, calculate the additional attenuation needed to meet the target level;

• Select and specify noise-reduction measures – use standard noise-control techniques (e.g., enclosures, attenuators, acoustic louvres, noise barriers, use of sound-absorbing panels, and vibration-isolation devices);

• Predict noise reductions to be achieved using selected noise-reduction measures and compare them with target levels; and

• If necessary, repeat the cycle until predictions indicate that noise limits will be met.

6.7 Effects on Health-Based Noise Limits

There is a trend towards basing noise limits on research into the effects of noise on health. The *WHO Night Noise Guidelines for Europe 2009* first introduced the concept of observable noise levels: the no observable effects level (NOEL); the lowest observed adverse effect level (LOAEL); the significant observed adverse effect level (SOAEL). These concepts have been incorporated into the *Noise Policy Statement for England* (NPSE) issued in 2010. This has a long-term policy vision of promoting good health and good quality of life through the effective management of noise within the context of government policy on sustainable development. The aims of the policy are to avoid significant adverse impacts on health and quality of life; to mitigate and minimise adverse impacts on health and quality of life; to contribute to the improvement of health and quality of life. For example, following research on the health effects of noise reported in *WHO Night Noise Guidelines for Europe 2009* for Europe, a night-time outdoor LAeq,T at 8 h of 40 dB could be adopted as a NOEL limit.

6.8 Extracts from BS 8233:2014 Relating to Industrial Noise

6.8.1 General

Industrial noise can originate from specific processes (either internal or external) to buildings, or from related transport operations, such as loading/unloading of vehicles or activities involving other plants, such as fork lift trucks.

6.8.2 Noise Emitted by Factories

If a proposed factory development is to be situated in the vicinity of noise-sensitive buildings, the local planning authority usually sets planning conditions that take account of any predicted increase in noise due to the factory (see Clause 5). Extensive noise-control measures might be required, especially if the noise is impulsive, has a strong tonal character, or is otherwise of a distinguishable nature.

On an industrial estate, the noisier factories should be sited furthest from houses, with warehouses and quieter production areas used as buffers between the noisier factories and dwellings outside the industrial estate. Careful site planning can give some protection to noise-sensitive activities on the estate.

Common causes of complaint, which should be taken into consideration, are noise from: industrial processes; external generators; calling systems; end-of-shift indicators; vehicle movements; night-time working.

6.9 Summary of Main Points

The main points covered in this chapter are summarised below.

1. Noise which may cause disturbance and nuisance to neighbours may originate from machines and processes within workplace buildings, from external sources such as generators, pumps, motors, HVAC equipment and from transport of material to and from the workplace, and associated loading and unloading of vehicles, and from sound from public-address systems and end-of-shift signals.

2. Noise from industrial sites affecting nearby communities will be subject to control by nuisance and planning legislation.

    In addition, certain industrial processes are required to meet permitted noise levels specified under IPPC regulations.

3. It is usual to specify noise limits outside noise-affected buildings rather than inside them because it is much easier to verify compliance by carrying out noise measurements outside.

4. Noise limits will usually be specified in terms of an LAeq,T value, supplemented by the maximum noise level, to deal with the effect of a short-duration burst of high-level noise (LAmax values), especially at night.

5. BS 4142:2014 is the most commonly used method for impact assessment of industrial noise in the UK. It provides a method for assessing the impact of industrial sound on people in nearby buildings. The method is based on the difference between the existing ambient noise level at the noise-sensitive premises and the sound level arising from the introduction of the new sound source, called the specific sound level. The main features of the method are described.

6. BS 7445/ISO 1996 is a broadly similar noise-impact assessment method to BS 4142:2014 in as much as both are based on the extent that a specific noise, measured as an LAeq,T value, exceeds existing ambient noise levels, but is much wider in scope, covering all types of environmental noise. BS 7445/ISO 1996 gives general guidance on noise measurement and assessment procedures and on the setting of noise limits but, unlike BS 4142:2014, does not suggest noise limits.

7. Fixed noise level criteria: WHO guidelines and BS 8233:2014.

The *WHO Community Noise Guidelines*, published in 1989, recommends that noise levels outside residential buildings should not exceed 55 dBA in the daytime or 45 dBA at night. More recent WHO guidelines (*WHO Night Noise Guidelines for Europe 2009*) recommend working towards the lower limit of 40 dBA outside at night-time.

BS 8233:2014 advises that in general, for steady external noise sources, it is desirable that the internal ambient noise level in living/dining rooms does not exceed the guideline value of 35 dB.

8. To relate guideline limits for noise levels inside buildings to those for noise level limits outside buildings, it is necessary to estimate the sound insulation provided by the building façade. A much used rule of thumb is that for residential dwellings with windows open, this is between 10 and 15 dBA, but BS 8233:2014 gives further guidance.

9. Good management operational measures are discussed to minimise noise disturbance to neighbours, including: location of noise-making plant and equipment; avoiding noisy operating at sensitive times; using quieter equipment if possible; use of noise shields, screens enclosures, maintenance of equipment and of buildings; operating equipment at low power and speeds where possible; planning and prediction to use appropriate noise-reduction measures.

10. Planning and prediction should be used to anticipate community noise problems and to put noise-control measures in place to prevent them happening. Noise levels outside noise-sensitive properties should be predicted and compared with specified noise limits and, where necessary, noise-attenuating measures put in place to reduce noise levels to the required target levels.

11. Maintaining good relations and good communications with noise-sensitive neighbours can help reduce levels of dissatisfaction and complaint. An effective noise-complaints procedure should be in place, with the appropriate staff trained to deal with complaints about noise effectively and sympathetically.

12. Health-based noise criteria: NOEL, LOAEL and SOAEL.

The *WHO Night Noise Guidelines for Europe 2009* introduced the concept of health-based 'observable effect (noise) levels': NOEL, LOAEL and the SOAEL. These concepts have been incorporated into the NPSE issued in 2010 which has a long-term policy vision of promoting good health and a good quality of life through the effective management of noise within the context of Government policy on sustainable development.

13. The chapter concludes with some extracts from BS 8233:2014 relating to industrial noise.

Bibliography

1. BS 4142:2014 – Methods for rating and assessing industrial and commercial sound.

2. BS 7445-1:2003 – Description and measurement of environmental noise – Guide to quantities and procedures.

3. BS 7445-2:1991, ISO 1996-2:1987 – Description and measurement of environmental noise – Guide to the acquisition of data pertinent to land use.

4. BS 7445-3:1991, ISO 1996-3:1987 – Description and measurement of environmental noise –Guide to application to noise limits.

5. BS 8233: 2014 – Guidance on sound insulation and noise reduction for buildings.

6. *Guidelines for Community Noise*, World Health Organisation, Geneva, Switzerland, 1989.
   Night Noise Guidelines for Europe, World Health Organisation, Geneva, Switzerland, 2009.

7. *Guidelines for Environmental Noise Impact Assessment*, Version 1.2, Institute of Environmental Management and Assessment, Lincoln, UK, November 2014.

8. Integrated Pollution Prevention and Control (IPPC) Regulations: Horizontal Guidance Note IPPC H3 (Part 2) – Noise Assessment and Control, Environment Agency, 2002. (This document contains useful information on the measurement, prediction and assessment of noise, and on methods for noise control).

9. *Noise Policy Statement for England*, Department for Environment, Food & Rural Affairs, London, UK, 2010.

10. S. Mitchell in *Best Available Techniques for Control of Noise & Vibration*, R&D Technical Report P4-079/TR/1, Environment Agency, 2001. (The work for this report was undertaken for the Environment Agency by ERM and partly forms the basis for guidance, H3, Part 2. In particular, it contains a number of case studies relating to investigation and control of noise problems that do not appear in this guidance).

11. *WHO Night Noise Guidelines for Europe*, World Health Organisation, Geneva, Switzerland, 2009.

12. *WHO Community Noise Guidelines*, World Health Organisation, Geneva, Switzerland, 1989.

7 Prediction of Noise Levels

7.1 Introduction

The ability to predict noise levels can play an important part in the control of noise.

To be sure of complying with any noise limits, it will be necessary for machinery operators to be able to predict noise levels, both within the workplace and external to the workplace, at the nearest noise-sensitive properties. When planning noise-reduction measures, including alterations to the layout of the workplace, having the capability to predict levels from different options can allow noise-control measures to be optimised.

7.2 Prediction (Simple Treatment)

The starting point is knowledge of the sound power level of each noise source (L_w) (i.e., each noisy machine) in dB, octave-band values, and the distance of the reception point from each of the sources.

The prediction of noise from machinery at relatively short distances, of up to a few hundreds of metres, depends mainly on the: sound power level of the sound source; distance between the source and receiver; presence of sound-reflecting surfaces either close to the source or to receiver; presence of sound-blocking barriers which prevent a direct line of sight between the source and receiver; nature of the ground between the source and receiver. At greater distances, weather-related factors such as wind/temperature gradients, atmospheric scattering and temperature- and humidity-related sound absorption by the air must also be taken into account.

7.3 Distance from the Noise Source

The sound power emitted by the source becomes spread out over increasingly larger areas as the sound travels away from the source, rather like the way ripples spread over the surface of water if a small stone is dropped onto it. This leads to a reduction in sound pressure level(s) (L_p) with distance from the source. The sound pressure level at a distance, r, from the, L_w is given in **Equation 7.1**:

$$L_p = L_w - 20 \times \log(r) - 11 \qquad (7.1)$$

Table 7.1 shows the predicted sound pressure level at a range of distances from a source of sound power level of 100 dB.

Table 7.1 Predicted sound pressure levels at a range of distances from a source of sound with a power level of 100 dB	
Distance in metres from a sound source with L_W = 100 dB	Sound-pressure level, L_p/dB
1	89
1.4	86
2	83
2.8	80
4	77
5	75
7	72
8	71
9	70
10	69
14	66
20	63
28	60
40	57
50	55
70	52
80	51
90	50
100	49
140	46
200	43
280	40
400	37
500	35
700	32
800	31
900	30
1,000	29

It can be seen that, for every doubling of distance, there is a 6 dB reduction in sound pressure level (e.g., from 2 to 4 m or from 4 to 8 m). Also, for every 10 fold increase in distance, there is a 20 dB reduction in sound pressure levels (e.g., from 1 to 10 m or from 10 to 100 m).

The effect of a different sound power level produces a corresponding similar change in sound pressure level, at the same distance (e.g., reducing the L_w value by 10 dB will cause a similar reduction of 10 dB in sound pressure level).

Thus, for example, to estimate the noise level at a distance of 10 m from a machine with a sound power level of 110 dBA would be 79 dBA (69 dBA from **Table 7.1** plus 10 dB for the additional 10 dB of sound power level).

7.3.1 Nearby Sound-Reflecting Surfaces

The formula stated above was based on no reflecting surfaces close to the sound source. This in effect assumes that the ground beneath the source is sound absorbing and not sound reflecting – in effect soft ground, such as grassland, soil, or carpeting, indoors. If the ground beneath the source is sound reflecting (i.e., hard ground such as tarmac or concrete), then the effect of the sound reflections will be to increase the sound levels predicted by the formula given above, or given in the form shown in **Table 7.1** by 3 dB.

If the source was located close to a second-reflecting surface, for example, at the base of a brick wall or a building façade, in addition to being on hard ground beneath then this would result in another 3 dB increase: a total increase of 6 dB over the values predicted by **Equation 7.1** or **Table 7.1**.

In a similar way, sound levels are increased by the presence of sound-reflecting surfaces close to the receiver.

7.3.2 Sound Levels in Indoor Spaces

Outdoors, if there were no reflecting surfaces, the sound received directly from the noise source would continue to reduce in level with increasing distance from the noise sources by, in theory, 6 dB for every doubling of distance until the background noise level was reached.

Inside a building this direct sound level will be augmented by sound reflected from the walls, floor and ceiling, and from other surfaces, such as furniture, or machinery.

The sum total (i.e., the accumulated effect of the very many reflections that will occur) is called 'reverberation', or 'reverberant sound'. The level of this reverberant sound depends on the size of the room and on the amount of sound absorption of the room surfaces, which reduces the amount of reflected sound (**Chapter 1**).

The reverberant sound builds up to a level where the rate at which sound is being supplied to the room by the source is equal to the rate at which it is being absorbed by the room surfaces during the multiple reflections.

7.3.3 Sound Distribution in Rooms – Direct and Reverberant Sound

A simple theory of reverberant sound assumes that the reverberant sound pressure level is constant throughout the room, and that there is no net flow of sound energy in any direction (i.e., that the sound is diffuse). The reverberant sound pressure level depends on the sound power level of the source and on the amount of acoustic absorption in the room, expressed as a room constant, R_c. It does not depend on the distance from the source.

The total sound level at any point in a room is the combination of the direct and reverberant sound levels. **Figure 7.1** shows how the sound pressure level in a typical room varies with the distance from the noise source in the room according to this simple theory. Close to the source, where the sound level falls with increasing distance, the direct sound is predominant. At greater distances the graph levels off because the reverberant sound becomes the major component of the total sound level. In this region, the sound level depends on the R_c, so that the same noise source can give rise to different sound pressure level, measured at the same distance, but in different rooms.

The simple theory of a constant reverberant level throughout the room seems to be fairly accurate for rooms in which the length, breadth and height are all of similar dimensions. In large open-plan offices, however, the length and breadth may be very large compared with the height. Under these circumstances, reflections from the ceiling and floor become relatively more important than from the walls. The simple theory then no longer holds true and the sound pressure level continues to fall with distance from the source, and does not 'level out as shown in **Figure 7.1**. This may also happen in factories, where there are also the added complications of unusual ceiling or roof constructions, and arrays of machines which emit, scatter and absorb sound.

The direct sound pressure level depends on the distance from the source, and on the sound power level and directivity of the noise source, and does not depend in any way upon the room.

More accurate (but also more complicated) noise level-prediction methods which are suitable for workplaces containing machinery are described in EN ISO 11690-3:1998.

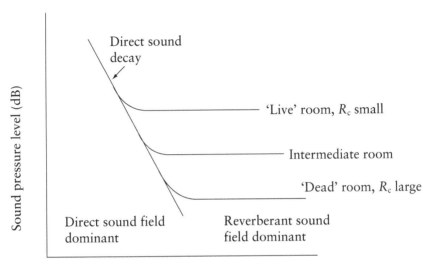

Figure 7.1 Variation with sound level *versus* distance from a source in a room. Reproduced with permission from R.J. Peters in *Noise Control (A Pira Environmental Guide)*, Pira, Leatherhead, UK, 2000. ©2000, Pira

7.3.4 Ray- or Beam-Tracing Methods

An alternative approach is to trace the path of rays (or narrow beams) of sound leaving the noise source taking into account the distance travelled and energy lost at each room surface. This involves creating a three-dimensional computer model of the workplace and assigning a sound-absorption coefficient (**Chapter 1**) to each surface, including machine surfaces. The sound power level in octave bands is also assigned to each noise source. The model tracks the path of each ray over a specified number of reflections (e.g., 50, 100, 200 or more) and the number of rays leaving each source is also specified. The total sound pressure from all of the rays at a specified receiver position, or an array of receiver positions, is calculated and a graphical display of noise level contours is presented.

This method of prediction is very common and several commercial packages are available, including Odeon, Catt, and EASE.

7.3.5 Outdoor Sound Propagation

When predicting sound levels outdoors, although the effect of distance from the source is the single most important factor to be considered, there are other factors which need to be taken into account, each of which will produce additional attenuation. The most important of these are:

- Ground attenuation.

- The effect of any barriers between the source and receiver.

- The effects of any reflecting surfaces close to the source or receiver.

- The bending of sound waves in the atmosphere (refraction).

- The absorption of sound in the atmosphere (which is likely to be significant only at long distances and high frequencies).

Each of these phenomena will be described briefly below.

7.3.6 Ground Attenuation

The nature of the ground between the source and receiver can also affect the sound level at the receiver, particularly if both the source and receiver are located at similar heights above, and both are close to ground level (within ≈1 m). The sound reaching the receiver will have arrived by two different paths, of similar lengths – directly from the source, and after reflection at the ground at a point half way between source and receiver. For some types of soft ground, and at certain sound frequencies the reflection leads to a change in both the amplitude and phase of the reflected sound wave. This results in a partial cancellation of the direct airborne sound wave by the reflected wave, and a total reduction or attenuation in the sound level at the receiver. This process is called 'ground attenuation' and can produce significant sound reductions at distances of a few tens of metres, particularly if both the source and receiver are located at similar heights above, and both close to ground level. This is often the case when the sound source is within a metre or so above ground level and the receiver is a standing or sitting person on the ground floor of a building. The ground attenuation affect would be much less, from the same sound source, for a receiver located on the upper floors of a multi-storey building.

For a source-to-receiver distance of 150 m and an average source and receiver height of 1.5 m, the predicted soft-ground attenuation is about 4 dBA, and as high as 25 dBA for distances of 1,000 m.

7.3.7 Effect of Barriers between the Source and Receiver

A solid barrier such as a wall or fence which prevents a direct line of sight between the sound source and receiver will also prevent direct sound transmission and so reduce the sound reaching the receiver provided that it is heavy enough and does not contain any holes or gaps. The only other way that sound from the source can then reach the receiver is by bending over the top and around the sides of the barrier. The amount of such bending (or 'diffraction' as it is called) depends on the size of the barrier and the wavelength of the sound, which in turn depends on the frequency of the sound.

This means that any noise barrier will be more effective in reducing high frequencies than for at low frequencies.

The effectiveness of a barrier in reducing sound transmission between the source and receiver is limited by the diffraction of sound over the top and around the sides of the barrier, which is determined by the size of the barrier compared with the wavelength of the sound. The attenuation provided by the barrier (also known as the 'insertion loss') is the reduction in noise level at the receiver arising from the noise source as a result of the presence of the barrier. The usual basis of prediction is that the sound transmission through the barrier is negligible and can the ignored compared with diffraction around the edges of the barrier. There are various methods for predicting the barrier attenuation arising from diffraction, but one of the simplest, described below for a long thin barrier, developed by Maekawa, is based on path difference.

The path difference, δ, [i.e., the additional sound path for the rays travelling over a simple barrier (**Figure 7.2**)] is given in **Equation 7.2**:

$$\delta = (a + b) - c \qquad (7.2)$$

The value of path difference, δ, may be obtained either from an accurate scale drawing of the situation or by calculation using trigonometry, and the Pythagoras theorem in particular, as shown in **Figure 7.2** below.

The table in **Figure 7.2** shows the value of insertion loss (in dB) for various values of path difference, and frequencies, in octave bands – for example, for a path difference of 0.2 at 250 Hz, the insertion loss in 9 dB.

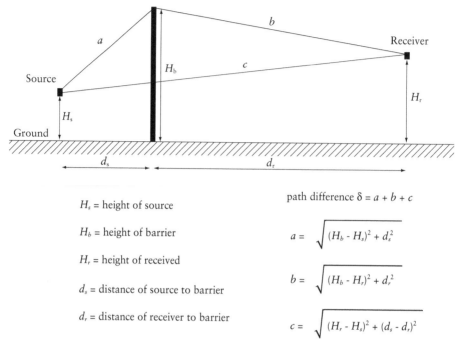

H_s = height of source

H_b = height of barrier

H_r = height of received

d_s = distance of source to barrier

d_r = distance of receiver to barrier

path difference $\delta = a + b + c$

$$a = \sqrt{(H_b - H_s)^2 + d_s^2}$$

$$b = \sqrt{(H_b - H_r)^2 + d_r^2}$$

$$c = \sqrt{(H_r - H_s)^2 + (d_s - d_r)^2}$$

δ	63	125	250	500	1,000	2,000	4,000
0	5	5	5	5	5	5	5
0.05	5	6	6	8	9	12	14
0.1	6	6	8	9	12	14	17
0.2	7	8	9	12	14	17	20
0.3	7	9	11	13	16	19	22
0.4	8	9	12	14	17	20	23
0.5	8	10	12	15	18	21	24
0.6	9	11	13	16	19	22	25
0.7	9	11	14	16	19	22	25
0.8	9	12	14	17	20	23	26
0.9	10	12	15	17	20	23	26
1	10	12	15	18	21	24	27

Figure 7.2 Calculation of barrier insertion loss. Reproduced with permission from R.J. Peters in *Noise Control (A Pira Environmental Guide)*, Pira, Leatherhead, UK, 2000. ©2000, Pira

In practical terms, the effectiveness of a noise barrier will be limited to between 10 and 20 dBA outdoors because of atmospheric scattering and diffraction of noise over the top of the barrier, and because of loss of ground attenuation if the barrier is on soft sound-absorbing ground.

The same barrier used indoors would be less effective because sound may be reflected around and over the top of the barrier by room surfaces. The effectiveness of noise barriers indoors may be increased by lining these nearby surfaces with sound-absorbing material to reduce sound being reflected around and over the top of the barrier, but will be limited in many cases to between 5 and 10 dBA.

7.3.8 Refraction of Sound in the Atmosphere

There is usually a gradual change of the speed of sound in air with the height above the ground. This causes the sound waves to change direction slightly: downwards (towards the ground), causing the sound level at ground level at a distance from the source to be higher; or upwards (away from the ground), causing the sound heard at ground level to be lower in level than otherwise. This bending of sound rays is called 'refraction', and the slight (but progressive) changes of sound level with height are caused by changes of air temperature and wind speed, both of which change with height above the ground depending on weather conditions.

On a dry day with clear skies, the air is warmest at ground level and the temperature decreases (causing the sound speed also to decrease) with height above the ground. Under these conditions, sound waves from a source to ground level bend slightly upwards towards the sky. Under these conditions, smoke from chimneys or bonfires rises vertically and, like sound, disperses with distance.

On overcast days, including when it is foggy or raining, the reverse can happen and air temperature can increase with height above the ground. These conditions are called 'temperature inversions', and cause sound to bend downwards toward the ground. The sound waves do not disperse with distance as well as they do on clear days, and so distant sounds appear to be louder. Under inversion conditions, smoke from chimneys does not disperse as well as on clear days and sometimes the effect can be seen when the upward-rising smoke meets an inversion layer of air and becomes trapped below the inversion.

Wind speed is usually lowest at ground level because of the effects of friction with the ground, and increases with height above the ground. The speed of sound is affected depending on whether the sound is travelling upwind or downwind. The changes of wind speed (and hence changes of sound speed) with height causes sound downwind of the source to bend downwards and conversely to bend upwards at locations upwind of the source. Thus, it is the common experience that distant sounds are louder when carried by the wind (i.e., downwind of the source) than when travelling against the wind (i.e., upwind of the source).

The effects of wind and temperature gradients on outdoor propagation are shown in **Figure 7.3.**

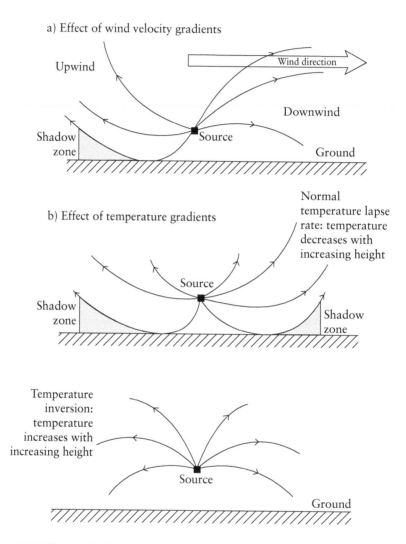

Figure 7.3 Effects of wind and temperature gradients on sound propagation. Reproduced with permission from R.J. Peters in *Noise Control (A Pira Environmental Guide)*, Pira, Leatherhead, UK, 2000. ©2000, Pira

It is the combined effects of wind and temperature changes in the atmosphere that account for the fact that on some days sounds from distant sources such as road traffic, church bells, sporting and entertainment venues, appear much louder than on other days.

This behaviour can sometimes become more complicated than described above, thereby allowing sound to disperse normally up to a certain height and then become trapped in within an inversion layer, before eventually bending downwards and returning to ground level. This can lead to unusual sound propagation effects whereby an event such as an explosion can be heard more loudly at distant positions than as distances closer to the source.

The effects of refraction of sound in the atmosphere are not usually significant at short distances of up to a few hundred metres, but for longer distances can become a major source of day-to-day fluctuations in sound levels from distant sound sources. The bending of sound effects can also cause sound to bend into the shadow zones of barriers between the source and receiver and have the effect of reducing the effectiveness of the barrier in terms of reducing sound levels. A similar effect can be caused by air turbulence when turbulent pockets of air 'scatter' sound over the top of barriers towards the receiver.

7.3.9 Absorption of Sound in Air

'Sound absorption' is the process whereby sound energy is converted to heat energy as a result of some frictional process affecting the vibratory motion of air particles which are the cause of the sound. Most sound absorption usually occurs when sound waves travelling in air meet sound-absorbing surfaces. The most effective sound-absorbing surfaces are porous and the sound absorption occurs as a result of friction between the vibrating air particles and the surface of the porous material.

However, some absorption occurs as the sound travels through the air itself, away from any absorbing surfaces, as a result of the viscosity of the air, thermal effects and from the vibration and rotation of air molecules.

Under normal circumstances, the absorption of sound by air is not considered very important when compared with other factors involved in sound propagation, such as the spreading of the sound wave-front, ground attenuation, refraction and scattering in the atmosphere, or diffraction by barriers. Air absorption becomes significant only when considering propagation outdoors over long distances (more than a few hundred metres), and particularly if high-frequency sound is involved. Air absorption is much more important at ultrasonic frequencies and is the reason why many technical devices (e.g., burglar alarms) employing ultrasonic beams have a limited range in air. Air absorption of sound energy becomes significant only in sound propagation indoors at the very highest frequencies and for the very largest spaces such as theatres and concert halls. Usually it is negligible compared with the absorption that occurs when the sound strikes a surface such as a wall or floor.

An approximate expression for the excess attenuation given by air absorption is given in ISO 9613 – Acoustics – Attenuation of sound during propagation outdoors – Part 2: A general method of calculation. At 20 °C and a relative humidity of 50% the predicted attenuation is:

- 0.5 dB/km at 500 Hz

- 1.5 dB/km at 1 kHz

- 6 dB/km at 4 kHz

7.4 Predicting Sound Levels Outdoors

As stated above, the starting point is to predict the sound pressure level at the required distance (i.e., at the nearest noise-sensitive property) from a sound source of known sound power level. The next step is to add correction factors (in dB) for the various other factors, the most important of which will be:

- Ground attenuation.

- The effect of any barriers between the source and receiver.

- The effects of any reflecting surfaces close to the source or receiver.

- The bending of sound waves in the atmosphere (refraction).

- The absorption of sound in the atmosphere (which is likely to be significant only at long distances and high frequencies).

These various effects will all vary with frequency so it will be necessary to carry out these additional attenuation calculations for each octave band, usually between 125 and 4,000 Hz, (but sometimes between 63 and 8,000 Hz) and then, if required, combine the sound pressure levels in each octave band to find the overall dBA or dBZ level.

If there is more than one source, at a different distance from the reception point (the nearest noise-sensitive property) then the process is repeated and the sound levels from each source are combined (using dB addition) to obtain the total predicted sound pressure level, from all sources at the single reception point.

If the noise level is required at various points at noise-sensitive properties, the process is repeated for each reception point.

If the average LAeq,T value is required over a period of time, for example, over a 16 h daytime period and the different sources contribute for different periods, then the individual LAeq,T values from each source will have to be calculated and then combined for all the sources combined.

Noise-level predictions will usually be made for each octave band. Hence, the final stage of the prediction will have to be to convert the octave bands to a single dBA noise level.

7.4.1 Calculations of Sound Level

Further information, including worked examples of outdoor sound-level prediction, may be found in textbooks, and also in various reports and standards (please see the bibliography at the end of this chapter).

Simple situations can be calculated using a spreadsheet but several 'noise mapping' software packages are available which will perform the calculations for more complex situations involving multiple sources and reception points and display the results graphically as colour-coded noise-level 'contour maps'. Examples of such packages include: CadnA, SoundPlan, IMMI, among others. Some of these packages can be adapted for indoor as well as outdoor noise modelling.

7.4.2 Examples of Noise-Level Predictions using Simple Acoustic Models

Although formulae and algorithms are given in various standards, guidance documents and textbooks for noise-level predictions using a calculator or spreadsheets for fairly simple cases, computerised software packages are available and commonly used for the prediction of noise level indoors and outdoors. These also have the advantage of built-in graphical displays, of noise level contours, for example, and enable the effects of 'what if' changes to be investigated quickly.

To illustrate the use of such software, two simple computer models have been created for an 'L-shaped' multi-storey commercial building: one for predicting internal noise levels and one for predicting the radiation of noise to the nearby external environment.

7.4.3 Internal Noise Model

The ground floor of the building is a production area with four main sources of noise; two compressors with sound power levels of 115 dBA, and two pumps with a lower sound power levels of 110 dBA. There are workstations where employees are situated throughout the area as well as at close to the four machines. The model has been used to investigate the possible benefit of separating noise and quiet areas. In the first version, **Figure 7.4a**, the two noisy machines are located at either end of the workspace with the two quieter machines in between. It can be seen that in the long arm of the 'L' the noise levels are dominated by the noise from the two noisy machines, and the lowest noise levels are between 80 and 85 dBA, although there are quieter areas around the corner in the toe of the 'L'.

In the second version (**Figure 7.4b**), the two noisy machines have been moved to be close together in the toe of the 'L' and the walls and ceiling in this area only have been modelled as being covered with sound-absorbing material to reduce reflected sound. The two quieter sources have also been located close together.

The result is that noise levels are much lower over much of the workplace area, although high noise levels still remain in the 'machine bays' in the toe of the 'L'. It will be necessary to issue employees who work in this area with suitable and effective hearing protection (**Chapter 5**) and, if possible, automate the operation of these two machines so that periods of attendance by employees is minimised.

Many further options can easily be tested using the model, such as: the effects of reducing machine sound power levels through noise-control treatments, or replacement by quieter machines; moving the machines to different positions; the use of noise screens or barriers; use of different sound-absorbing treatments to walls and ceiling areas.

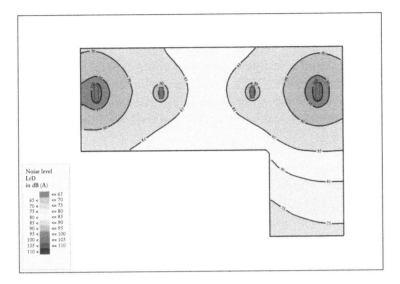

a)

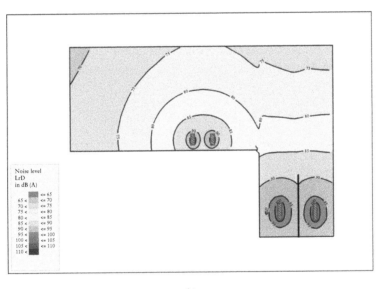

b)

Figure 7.4 Internal noise models illustrating the effects of separating noisy and quieter sources (a) quiet and noisy sources not separated and (b) the two noisy sources grouped together, with additional sound absorption to walls and ceiling close to the noisy machines (LrD: noise level at the receiver in the daytime). Reproduced with permission from R. Tipping and J. Richards of KBR Ltd. ©KBR Ltd

7.4.4 External Noise Model

The L-shaped multi-storey commercial building has four external noise sources which radiate noise to the surrounding area: a rooftop plant source (e.g., a chiller), a ground-level outdoor source (e.g., a compressor), noise radiating from louvres in one face of an attached single-storey boiler house/plant room, and a tall extract stack adjacent to the plant room, radiating noise from the top of the stack.

There is a block of flats nearby, with the highest floors having a direct line of sight of the rooftop plant, but the lower floors do not have a line of sight. There are also some houses nearby. There are also some single-storey buildings (e.g., garages) in between the commercial building and the flats and houses. The external ground-level noise source is on the same aside of the building as the noise-sensitive properties. The surrounding ground is flat and hard and so assumed to be sound reflecting.

The scenario is shown in **Figures 7.5a** and **b.** It is possible to run the model with any combination of the four different noise sources in operation, and doing so indicates that it is the noise from the externally located generator that is the dominant contributor to the noise levels at the nearby noise-sensitive properties. **Figure 7.5c** and **d** show the effect of moving the generator to the side of the building away from the noise-sensitive properties, which reduces the noise levels at these properties by about 5 dBA, from 50 to 45 dBA.

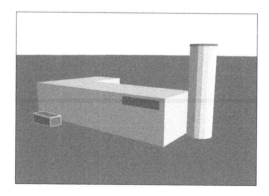

a)

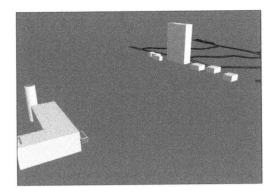

b)

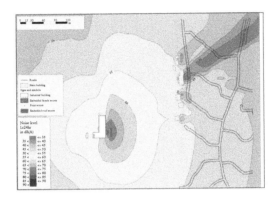

c)

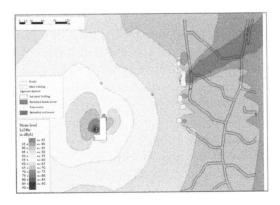

d)

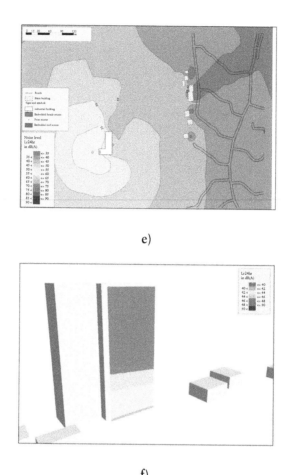

e)

f)

Figure 7.5 (a) Four external noise sources, from left to right: generator, rooftop chillers, louvre in plant room, and top of exhaust stack. (b) Building with four noise sources and nearby noise-sensitive properties. (c) All sources with a diesel generator set located to the east side of the building. (d) All sources with a diesel generator set located to the west of the building. (e) Roof source (chillers) only. (f) Roof sources (chillers) only in operation. Noise levels at the façade of the block of flats arising from the chillers at roof top level (Lr24hr: noise level at the receiver over 24 h). Reproduced with permission from R. Tipping and J. Richards of KBR Ltd. ©KBR Ltd

At night-time the only noise sources in operation are the chillers on the roof of the building. **Figure 7.5e** shows that the sound level at ground level at the flats and other buildings arising from the operation of the chillers is about 40 dBA. **Figure 7.5f** shows the noise levels at the vertical façade of the block of flats. The noise levels vary with the height above the ground from 40 to 48 dBA.

The effect of various noise-reducing measures to be investigated include reducing the sound power levels of the chillers through noise-control treatments such as fitting attenuators, enclosure or screens around the chillers, or replacement by a quieter version.

7.5 Summary

The ability to predict noise levels can play an important part in the control of noise by allowing noise-control measures, if needed, to be planned more effectively and economically in advance of the introduction of potentially noise developments.

The starting point is to know the sound power level of the sound sources. The first stage is to consider the way in which the sound level from the source varies with distance from the source, taking into account the effect of any nearby sound-reflecting surfaces.

In an indoor situation, this direct sound (i.e., sound travelling directly from the source to the receiver) is supplemented by multiple reflections from the room surfaces, called reverberant sound or reverberation. According to a simple theory, the direct sound reduces with distance from the source at a rate of 6 dB per doubling of distance, but the level of reverberant sound remains constant throughout the room and depends on the room size and the amount of sound absorption that it contains. The direct sound depends only on the distance from the source and not upon the room surfaces.

The total noise level at any point in the room is obtained by combining the direct and reverberant sound level. Close to the source the sound level is mainly determined by the direct sound from the source but, as the distance from the source increases, the reverberant sound level becomes more important.

The calculation of direct and reverberant sound levels can be carried out using a computer spreadsheet for fairly simple situations, but an alternative is to use computer ray- or beam-tracing software packages. This method of prediction is very common and several commercial packages are available, including Odeon, Catt, and EASE, among others.

Outdoors, without the influence of room reflections, there are several additional mechanisms which lead to the attenuation of sound in addition to the attenuation provided by an increasing distance from the source. These mechanisms include the effects of ground attenuation, barriers, refraction or bending of sound waves in the atmosphere, and the absorption of sound by the air. Simple situations can be calculated using a spreadsheet but several noise mapping software packages are available that will perform the calculations for more complex situations involving multiple sources and reception points and display the results graphically as colour-coded noise-level contour maps. Examples of such packages include CadnA, SoundPlan and IMMI.

This chapter concludes with an illustration using two simple software models for predicting noise levels: an indoor situation and an outdoor situation. The models illustrate how the ability to predict noise levels can help plan the arrangement of sound sources to minimise noise levels, and to investigate the effectiveness of various noise-control options.

Bibliography

1. R.J. Peters in *Noise Control (A Pira Environmental Guide)*, Pira, Leatherhead, UK, 2000.

2. I.J. Sharland in *Woods Practical Guide to Noise Control*, Fläkt Woods Limited (formerly Woods of Colchester), Colchester, UK, 1972.

3. *Noise Control in Industry*, Ed., J.D. Webb, Sound Research Laboratories, Little Waldingfield, UK, 1976.

4. *Noise Control in Mechanical Services*, Sound Research Laboratories, Little Waldingfield, UK, 1976.

5. R.J. Peters, B.J. Smith and M. Hollins in *Acoustics and Noise Control*, 3rd Edition, Prentice Hall, Upper Saddle River, NJ, USA, 2011.

6. D.A. Bies and C.H. Hansen in *Engineering Noise Control*, 2nd Edition, E & FN Spon, London, UK, 1996

7. ISO 9613 – Acoustics: Attenuation of sound during propagation outdoors – Part 2: A general method of calculation.

8. BS 5228-1:1997 – Noise and vibration control on construction and open sites. (Annex F of this standard provides methods for estimating the LAeq,T levels, taking into account: sound power outputs of processes and plant; periods of operation of processes and plant; distances from sources to receivers; presence of screening by barriers; reflection of sound; and soft ground attenuation. The levels from the range of equipment used are combined to give an overall LAeq,T level.)

9. IPPC H3 – Horizontal Guidance for Noise – Part 2: Noise assessment and control, 2002. (This report gives useful guidance on various aspects of noise measurement, methods of assessing noise levels, sound level calculations and prediction methods, and methods of noise control.)

10. K.J. Marsh, *Applied Acoustics*, 1982, **16**, 411.

11. EN ISO 11690-3:1998 – Acoustics – Recommended practice for the design of low-noise workplaces continuing machinery – Part 3: Sound propagation and noise prediction in workrooms.

8 Specification of Noise Emission from Machinery and Machinery Noise Regulations (The European Union Machinery Noise Regulations)

8.1 Introduction

According to a simple application of the inverse square law, a portable electricity generator with a sound power level of 110 dBA will create a sound pressure level (L_p) of 92 dBA at a distance of 5 m away outdoors on soft ground (grass) well away from any sound-reflecting surfaces. If placed on hard ground (e.g., tarmac or concrete), the sound pressure level from the same source at the same distance increases by 3 dB to 95 dBA because of the reflections of sound from the ground to the reception position. If placed at the base of a vertical wall on the same hard ground (i.e., now close to two reflecting surfaces) it produces another 3 dB increase. If the same machine is measured at a distance of 5 m in a room such as a workshop, the sound pressure level will now be increased because of the influence of the floors, walls and ceilings and any other sound-reflecting surfaces nearby, and the measured sound level will depend on the nature of the room, whether it is large or small and whether it is 'live' or 'dead' (i.e., depending on the amount of sound absorption in the room to control the reflections).

The point of this illustration is that the same machine will produce different sound pressure levels (and sound louder or quieter) at the same distance depending on the acoustic environment. However, the sound power level of the machine – the amount of sound energy it produces – has remained constant.

Therefore, the sound power level is the preferred method for specifying noise emission from machinery, although there are some cases where the sound pressure level produced at a specified distance is used – in which case it is necessary to describe precisely the environment in which the sound level measurements were taken (e.g., the distances from and nature of nearby sound-reflecting surfaces).

Sound power levels have the advantage that the noise-emission characteristics of a machine are uniquely specified in one piece of information. However, they have the disadvantage that the sound power levels are not as easy to measure nor as easy to understand as sound pressure levels which are easily measured directly with a sound-level meter at any distance required, and these levels soon become easily understood and related to what people hear. By contrast, the accurate determination of sound power levels requires sound levels from the machine to be measured in a specialist test environment (an anechoic or reverberant test room), although approximate values can be determined from sound level measurements made *in situ*.

The sound power level value remains an abstract number until, as a result of theoretically based calculations, it can be related to a sound pressure level at some distance from the machine.

In free-field tests (i.e., in an anechoic room where there will be no sound reflections) the sound power level is determined from sound level measurements taken over a specified test surface surrounding the machine. Sometimes, if an anechoic room is not available, approximate values of sound power level can be estimated from sound level measurements outdoors, well away from reflecting surfaces or in large indoor spaces, providing that background noise levels are low enough.

There are many detailed standard methods for determination of sound power level for different types of machinery. They mainly differ in the specification of the test surfaces, which tend to be hemispheres for small machines and rectangular box-shaped surfaces ('parallelepipeds') for large items of equipment.

Perhaps the most widely used standards are the BS EN ISO 3740:2001 series, the first of which contains guidance on the different methods (and which of the standards in the series) to be used in different circumstances. The series includes three methods with three grades of accuracy: precision methods (grade 1), engineering methods (grade 2) and survey methods (grade 3). Precision grade methods require laboratory standard test rooms (anechoic or reverberant); the two other methods may be carried out with the machine *in situ*.

In all cases it will be necessary to specify the details of the operating conditions of the machine (e.g., the way in which it is fixed or mounted, operating speed, load, type of workpiece or material being processed).

Different sectors of industry have published their own noise test codes for measurement and specification of noise emissions.

8.1.1 Why is the Specification of Noise Emission So Important?

The specification of noise emission is important because the best way to ensure a quieter workplace in the future is to buy quieter machines when existing ones are replaced. Clear specification of noise emission is, therefore, important for the potential purchaser and to provide incentive for the producer of the machinery to reduce noise emission as far as possible to be competitive.

There are also machinery noise regulations which apply to manufacturers and suppliers of machinery. These regulations require publication of similar noise-emission information and, in some cases, impose limits on noise-emission levels of some types of machinery.

8.1.2 Limitation of Sound Power Levels

Sound power levels cannot be used to predict noise levels very close to large machinery (e.g., at the ear positions of operators). Hence, noise test codes specify the position of maximum noise levels around machinery and/or at operators' ear positions should also be measured as well as the array of measurement positions for the determination of sound power levels.

8.2 European Union Machinery Noise Directives

There are two European Union (EU) directives currently in operation: the machinery directive (2006/42/EC); and the outdoor directive (2000/14/EC). Both of these originate from Article 114 of the EU relating to product requirements and the free movement of goods. In other words, these are designed to ensure that, as far as machinery noise emission is concerned, the EU requirements are the same throughout Europe. The machinery noise directive is mainly concerned with noise in the workplace and protecting the hearing of employees from damaging levels of noise exposure. The outdoor directive is related to minimising disturbance from noise from outdoor machinery in the environment.

Both directives require that noise emissions from machinery should be openly declared and that manufacturers should reduce the noise of their products by the use of noise-reduction techniques as far as possible. Hence, both directives are helpful to purchasers of equipment and machinery by informing them of noise-emission levels from different competing products, and it is the intention that this will be an incentive to manufactures to develop quieter machines to obtain competitive advantage.

8.2.1 Directive 2006/42/EC (Noise and Machinery Directive)

In the UK, the directive is implemented in the *Supply of Machinery (Safety) Regulations 2008* (Statutory Instrument No.1597, 2008).

The following information relating to noise emissions must be declared in sales literatures, handbooks, and instruction manuals:

- A-weighted emission sound pressure level (L_{pA}) at workstations, where this exceeds 70 dBA; where this level does not exceed 70 dBA, this must be stated;

- A-weighted sound power level (LwA) emitted by the machinery where the L_{pA} at workstations exceeds 80 dBA; and

- Peak C-weighted instantaneous sound pressure level at workstations, where this exceeds 63 Pa (130 dB relative to 20 µPa).

For very large machinery, instead of the sound power level, the equivalent continuous L_{pA} at specified positions around the machinery may be given. Often, this is at positions around the machine at a distance of 1 m from its surface.

Where workstations are not defined or cannot be defined, sound pressure levels must be measured at a distance of 1 m from the surface of the machinery and at a height of 1.6 m from the floor or access platform. The position and value of the maximum sound pressure must be indicated.

8.2.2 Directive 2000/14/EC (Noise Emission of Outdoor Equipment Directive)

This directive requires that the LwA be declared for all types of machinery, and upper limits are imposed for some types of machinery.

The Directive is implemented in the UK by *The Noise Emission in the Environment by Equipment for use Outdoors Regulations 2001* (S.I. 2001/1701).

8.2.3 Standards

The sound levels must be measured using the most appropriate method for the particular machinery. Test methods and operating conditions must be clearly stated. Test codes are being developed to define standard methods of test and machine mounting and operating conditions for various categories or types of machine.

The following standards are recommended for use in meeting the requirements of the directives:

- BS EN ISO 3740:2001 series for the determination of sound power levels from sound pressure levels measurements.

- ISO 11200:2014 series for the determination of emission sound pressure levels.

- BS EN ISO 4871:2009 on the declaration and verification of noise-emission values of machinery and equipment.

8.3 Noise Test Codes

Noise test codes are test methods, usually given in a European or International Standard, which define how to obtain values for the noise emission of a machine. The test code will define parameters such as the conditions under which the machine will operate during tests and noise-measurement locations.

8.3.1 Noise Test Codes for Particular Industries

These are usually published as ISO standards. Examples include:

- ISO 1680:2013 – Acoustics – Test code for the measurement of airborne noise emitted by rotating electrical machines.

- ISO 2151:2004 – Acoustics – Noise test code for compressors and vacuum pumps – Engineering method (Grade 2).

- ISO 4412-1:1991 – Hydraulic fluid power – Test code for determination of airborne noise levels – Part 1: Pumps and Part 2: Motors.

- ISO 230-5:2000 – Test code for machine tools – Part 5: Determination of the noise emission.

- Noise test codes for plastics and rubber granulators and shredders are included as Annex A of: BS EN 12012-1:2007+A1:2008 – Plastics and rubber machines – Size reduction machines:

  BS EN 12012-1:2007+A1:2008 – Part 1: Safety requirements for blade granulators, Annex A: Noise test code.

  BS EN 12012-1:2007+A1:2008 – Part 3: Safety requirements for shredders, Annex A: Measurement and declaration of noise-emission values.

8.3.1.1 Part 1: Blade Granulators

BS EN 12012-1:2007+A1:2008 – Section 4.2 of the main part of the standard, is on hazards generated by noise warns that noise can cause: hearing damage; accidents due to interference with speech communication and interference with the perception of acoustic signals; physiological disorders.

BS EN 12012-1:2007+A1:2008 – Section 5.3.1 on noise reduction at source by design, states that *'when designing the machine, the available information and technical measures to control the noise at source shall be taken into account, see for example EN ISO 11688-1:1998'*.

NB: Useful information on noise generation mechanisms in machinery is given in EN ISO 11688-2:2000.

BS EN 12012-1:2007+A1:2008 – Section 5.3.2 on main sources of noise and noise reduction measures, states that the main sources of noise are the cutting chamber, hopper, feed opening, discharge opening, suction systems and discharge pipes, if provided.

Among the measures which may be taken are: changing the geometry of the blades and rotor; changing the geometry of the hopper; increasing the sound insulation of the cutting chamber; reduction of cutting speed; acoustic enclosures.

BS EN 12012-1:2007+A1:2008 – Section 5.3.3 on measurement and declaration of noise emissions, states that measurement and the declaration of noise emission shall be carried out according to Annex A.

BS EN 12012-1:2007+A1:2008 – Section 7.1.6, states that the manufacturer shall indicate (in the machine instruction manual) the operating conditions and the type of installation under which the noise test has been carried out and give information concerning noise-emission values from the blade granulator determined according to Annex A, and that (BS EN 12012-1:2007+A1:2008 – Section 7.1.7) that the manufacturer shall recommend that hearing and eye protection shall be used.

8.3.1.1.1 BS EN 12012-1:2007+A1:2008 – Annex A (Noise Test Code)

Noise-emission values may vary with machine operating conditions. Therefore, this noise test code specifies that noise-emission values shall be determined for at least two different test samples with the exception of blade granulators which are designed to granulate a specific item from a particular material.

It specifies noise-measurement methods and operating and mounting conditions that shall be used for the test.

8.3.1.1.2 Determination of Sound Power Levels

The A-weighted sound power level should be determined using either one of the BS EN ISO 740:1999 series (EN ISO 3741:1999; EN ISO 3743-1:1995; EN ISO 3743-2:1997; EN ISO 3744:1995; EN ISO 3745:2003; EN ISO 3746:1995; and EN ISO 3747:2000) which use sound pressure level measurements employing a sound-level meter, or using EN ISO 9614-1:1995 or EN ISO 9614-2:1996 which require the use of a sound-intensity meter.

An engineering method (grade 2 of accuracy) shall be used. If none of the engineering methods is practical, then a survey method (grade 3 of accuracy) shall be used. Reasons for choosing a survey method shall be given.

The duration of each measurement at each microphone position shall be at least 90 s. The measurement surface shall be a parallelepiped and the measurement distance shall be 1 m.

8.3.1.1.3 Determination of Noise Emission Sound Pressure Levels

For blade granulators that are so designed that they can be manually fed, the L_{pA} shall be measured using one of the standards EN ISO 11201:1995; EN ISO 11202:1995; or EN ISO 11204:1995.

8.3.1.1.4 Installation, Mounting and Operating Conditions

The blade granulator shall be mounted and connected as indicated by the manufacturer in the instruction manual of the machine.

Noise emissions from ancillary fixed power-operated feeding devices (e.g., conveyor belts) shall not be taken into account, but the noise emission of unloading equipment shall be taken into account.

New blades shall be used for the measurement. The feeding rate of the samples is determined by a formula given in the standard.

8.3.1.1.5 Information to be Recorded in a Test Report

General information: precise identification of the machine under test, mounting and operating conditions, acoustic environment, instrumentation and acoustical data.

Machine type, serial number, year of manufacture of the machine; feeding equipment; material temperature; ambient temperature; nominal horsepower of machine in kW; rotor details: diameter; rotation speed; number of blades; number of fixed blades; screen plate characteristics; rate of production of granulates in kg/h; airflow rate in m³/h when suction unloading equipment is used.

Test details: standards used for noise measurement, description of the mounting and operating conditions of the machinery during test, including details of the shape of samples used and of the physical characteristics of the materials used for the test.

8.3.1.1.6 Measurement Uncertainty

As with all measurements, there is an uncertainty associated with the noise-level measurements used to determine noise-emission values. The method for estimating this uncertainty is described in the code.

8.3.1.1.7 Declaration and Verification

The declaration and verification of noise-emission values shall be made in accordance with EN ISO 4871:1996.

NB (warning): The above summary of the noise test code has been included to be informative and helpful, but good professional practice requires that anyone required to use or apply the code and associated standards should read through the full versions of code and standards rather than relying on such summaries.

8.3.1.2 Part 3: Shredders

The noise test code for shredders is very similar to that for granulators. The main differences are as given below.

The shredder shall be placed on a plane-reflecting surface made of concrete. If elastic mounts are placed between the machine and the supporting surface, their technical characteristics shall be recorded.

Ancillary discharge equipment, whatever its nature, is not covered. Feeding devices (where feeding is automatic) are covered.

The machine shall operate at the rotor speed corresponding to the maximum nominal throughput, under 'no load' conditions. It is recommended that manufacturers start gathering noise-emission data under load conditions. These data can be obtained from measurements carried out by manufacturers at user workplaces on machines newly installed or during the installation phase of new machines.

Survey of Noise Emission and Risk Information Supplied with a Range of Work Machinery prepared in 2013 by the Health and Safety Laboratory for the Health and Safety Executive (HSE), Research Report RR962.

In a survey carried out of 73 sets of instructions obtained from the manufacturers and suppliers of 14 different categories of tools and machines, the noise-emission information contained in more than 60 of these instructions (82% of the sample) failed to satisfy the requirements of the machinery safety regulations. The failures related to: absent or incomplete declared noise-emission values; absent or incomplete traceability to operating conditions for declared noise-emission values; the lack of adequate (or in some cases any) information on safe use, residual risks, or noise-control measures

Several machine categories were identified during this study where further investigation of the noise emission and real-use noise exposure should be considered. These include granulators and shredders for processing of rubber and plastic materials:

- Granulators: granulators can generate very high noise levels, assuming the declared noise-emission values are credible. Based on anecdotal evidence of a growing supply of these machines from outside the EU, it is increasingly likely that users will have to rely on the noise-emission information provided by suppliers rather than manufacturers. Guidance should be provided to UK importers to help them influence manufacturers supplying into the UK or to ensure they take responsibility for providing adequate noise-emission data to users.

- Shredders: shredders can generate very high noise levels. The safety standard for shredders seems to suggest an additional problem with this category of machine is that they can only operate at the manufacturer's premises under no load; this claim seems worthy of further investigation. The safety standard recommends that manufacturers start gathering noise-emission data under load at user premises on newly installed machines or during the installation of new machines. There is no evidence that this is happening in the instructions reviewed during this study. It is also noted that the supply situation for shredders is similar to that described for granulators.

The report includes the following comments on the noise test codes given in **Table 8.1**:

Table 8.1 Summary of comments on the noise test codes	
Machine type	Comments
Plastics and rubber machines – blade granulators	BS EN 12012-1:2007+A1:2008 This standard applies to blade granulators, which are used to reduce objects and materials made from plastics and rubber into granules. The main sources of noise are identified (e.g., cutting chamber, hopper, feed opening) and details are given of some of the measures for reducing noise (e.g., increasing the sound insulation of cutting chambers, reduction of cutting speed, acoustic enclosures). The standard includes a test code, with methods for measuring sound pressure levels and sound power levels and operating conditions. This machine has been included in this project because the test code seems credible (i.e., noise-emission data are obtained under real conditions with no option to test the machine under no load). These machines are widely used, and some of the larger ones are likely to be noisy. It will be interesting to see whether manufacturers are following the specified test code.
Plastics and rubber machine shredders	BS EN 12012-3:2001+A1:2008 This standard covers shredders used for plastics and rubber. The machine begins with the outer edge of the feed hopper and ends with the discharge area. The main sources of noise are identified (e.g., shredding chamber, hopper, feed opening) and details are given of some of the measures for reducing noise (e.g., reduction of cutting speed, acoustic enclosures – if possible the chamber should be fitted with silencing in the outlet or discharge area so that structure-borne noise is isolated). The standard includes a test code specifying methods for measuring both sound pressure levels and sound power levels. The operating conditions are that the machine shall operate with no load at the rotor speed corresponding to the maximum nominal throughput. The reason these conditions are specified is that shredders can only operate at the manufacturer's place with no load and manufacturers currently do not measure noise at the user's place when installing new machines. It is recognised that noise emission under no load conditions is not representative of the noise emitted in normal operation. It is, therefore, recommended that the manufacturers start gathering noise-emission data under load. These data can be obtained from measurements carried out at user places on machines newly installed or during the installation phase of new machines. This machine has been included to investigate which noise-emission data are available from shredder manufacturers.
Reproduced with permission from J. Patel in *Survey of Noise Emission and Risk Information Supplied with a Range of Work Machinery*, Research Report RR962, Health and Safety Executive, Bootle, Merseyside, UK, 2013. ©2013, HSE	

8.4 Summary

The main points in this chapter are summarised below.

1. The noise energy emitted by a machine (i.e., its noise-emission level) may be specified either in terms of the sound pressure arising from the source at a specified distance or in terms of the sound power level of the source.

2. However, the same machine will produce different sound pressure levels (and sound louder or quieter) at the same distance depending on the acoustic environment – but the sound power level of the machine – the amount of sound energy it produces, has remained constant.

3. The sound power level is the preferred method for specifying noise emission from machinery, although there are some cases where the sound pressure level produced at a specified distance is used.

4. Sound power levels have the advantage that the noise-emission characteristics of a machine are uniquely specified in one piece of information. However, they have the disadvantage that the sound power levels are not as easy to measure nor as easy to understand as sound pressure levels which are easily measured directly with a sound-level meter at any distance required, and these levels soon become easily understood, and related to what people hear.

5. The specification of noise emission of machinery is important because the best way to ensure a quieter workplace in the future is to buy quieter machines when existing ones are replaced (the *Buy Quiet* policy). Therefore, clear specification of noise emission is important for the potential purchaser, and also to provide an incentive for the producer of machinery to reduce noise emission, as far as possible, to be competitive.

6. There are two EU directives on noise from machinery currently in operation: the machinery directive (2006/42/EC); and the outdoor directive (2000/14/EC). The machinery noise directive is mainly concerned with noise in the workplace and protecting the hearing of employees from damaging levels of noise exposure. The outdoor directive is related to minimising disturbance from noise from outdoor machinery in the environment.

7. Brief details of the requirements of these directives are given, including lists of relevant standards for determination of sound power levels and noise emission sound pressure levels.

8. Noise test codes are test methods, usually given in a European or International Standard, which define how to obtain values for the noise emission of a machine. The test code will define parameters such as the conditions under which the machine will operate during tests and noise-measurement locations.

9. Different sectors of industry have published their own noise test codes for the measurement and specification of noise emissions. Noise test codes for plastics granulators and shredders are described briefly.

10. A report commissioned by the HSE indicates that information provided by machine manufacturers about the noise emission of their products is often less than satisfactory and fails to meet the requirement of the machinery directives.

References

1. BS EN 12012-1:2007+A1:2008 – Plastics and rubber machines – Size reduction machines – Part 1: Safety requirements for blade granulators.

2. BS EN 12012-3:2001+A1:2008 – Rubber and plastics machines – Size reduction machines – Part 3: Safety requirements for shredders.

3. BS EN ISO 12001:2009 – Acoustics – Noise emitted by machinery and equipment – Rules for the drafting and presentation of a noise test code.

4. BS EN ISO 11689:1996 – Acoustics – Procedure for the comparison of noise-emission data for machinery and equipment.

5. J. Patel in *Survey of Noise Emission and Risk Information Supplied with a Range of Work Machinery*, Research Report RR962, Health and Safety Executive, Bootle, Merseyside, UK, 2013.

6. ISO 11201:2010 – Acoustics – Noise emitted by machinery and equipment – Determination of emission sound pressure levels at a work station and at other specified positions in an essentially free field over a reflecting plane with negligible environmental corrections.

7. ISO 11202:2010 – Acoustics – Noise emitted by machinery and equipment – Determination of emission sound pressure levels at a work station and at other specified positions applying approximate environmental corrections.

8. ISO 11204:2010 – Acoustics – Noise emitted by machinery and equipment – Determination of emission sound pressure levels at a work station and at other specified positions applying accurate environmental corrections.

9.   EN ISO 4871:1996 – Acoustics – Declaration and verification of noise emission values of machinery and equipment.

10. *Supply of Machinery (Safety) Regulations 2008*, Statutory Instruments No.1597, UK, 2008.

11. *The Noise Emission in the Environment by Equipment for use Outdoors Regulations 2001'* (S.I. 2001/1701), Health and Safety Executive, Bootle, Merseyside, UK, 2001.

12. *'Buy Quiet'*, Health and Safety Executive, Bootle, Merseyside, UK.

13. BS EN ISO 3740:2001 – Acoustics – Determination of sound power levels of noise sources – Guidelines for the use of basic standards.

14. 2006/42/EC – New machinery directive.

15. 2000/14/EC – Noise emission in the environment directive.

16. ISO 11200:2014 – Acoustics – Noise emitted by machinery and equipment – Guidelines for the use of basic standards for the determination of emission sound pressure levels at a work station and at other specified positions.

17. BS EN ISO 4871:2009 – Acoustics – Declaration and verification of noise emission values of machinery and equipment.

18. ISO 1680:2013 – Acoustics – Test code for the measurement of airborne noise emitted by rotating electrical machines.

19. ISO 2151:2004 – Acoustics – Noise test code for compressors and vacuum pumps – Engineering method (Grade 2).

20. ISO 4412-1:1991 – Hydraulic fluid power – Test code for determination of airborne noise levels – Part 1: Pumps.

21. ISO 230-5:2000 – Test code for machine tools – Part 5: Determination of the noise emission.

22. EN ISO 11688-1:1998 – Acoustics – Recommended practice for the design of low-noise machinery and equipment – Planning.

23. EN ISO 11688-2:2000 – Acoustics – Recommended practice for the design of low-noise machinery and equipment – Introduction to the physics of low-noise design.

24. BS EN ISO 740:1999 – Anaesthetic workstations and their modules – Particular requirements.

25. EN ISO 3741:1999 – Acoustics – Determination of sound power levels of noise sources using sound pressure – Precision methods for reverberation rooms.

26. EN ISO 3743-1:1995 – Acoustics – Determination of sound power levels of noise sources – Engineering methods for small, moveable sources in reverberant fields – Comparison for hard-walled test rooms.

27. EN ISO 3743-2:1997 – Acoustics – Determination of sound power levels of noise sources – Engineering methods for small, movable sources in reverberant fields – Methods for special reverberation test rooms.

28. EN ISO 3744:1995 – Acoustics – Determination of sound power levels of noise sources using sound pressure – Engineering method in an essentially free field over a reflecting plane.

29. EN ISO 3745:2003 – Acoustics – Determination of sound power levels of noise sources using sound pressure – Precision methods for anechoic and hemi-anechoic rooms.

30. EN ISO 3746:1995 – Acoustics – Determination of sound power levels of noise sources using sound pressure – Survey method using an enveloping measurement surface over a reflecting plane.

31. EN ISO 3747:2000 – Acoustics – Determination of sound power levels of noise sources using sound pressure – Comparison method in situ.

32. EN ISO 9614-1:1995 – Acoustics – Determination of sound power levels of noise sources using sound intensity – Measurement at discrete points.

33. EN ISO 9614-2:1996 – Acoustics – Determination of sound power levels of noise sources using sound intensity – Part 2: Measurement by scanning.

34. EN ISO 11201:1995 – Acoustics – Noise emitted by machinery and equipment – Measurement of emission sound pressure levels at a work station and at other specified positions – Engineering method in an essentially free field over a reflecting plane.

35. EN ISO 11202:1995 – Acoustics – Noise emitted by machinery and equipment – Measurement of emission sound pressure levels at a work station and at other specified positions – Survey method in situ.

36. EN ISO 11204:1995 – Acoustics – Noise emitted by machinery and equipment – Measurement of emission sound pressure levels at a work station and at other specified positions – Method requiring environmental corrections.

9 Towards a Quieter Workplace

9.1 Introduction

Although the priority should always be given to using quieter machinery and equipment wherever possible, much can also be done to reduce noise levels by careful design of the workplace. This short chapter gathers together and discusses these various strategies.

An employee at a workstation with a direct line of sight of a noisy machine will receive noise directly from the machine ('direct noise') and also from reflected and scattered sound arising from the same machine ('reverberant noise'). There may be several employees, each receiving both direct and reverberant sound from several different machines.

The direct sound may be reduced by increasing the distance between the source (the noisy machine) and the receiver, by interposing a screen partition between the source and receiver, or by an enclosure around the machine. The level of reflected (or reverberant) sound may be reduced by sound-absorbing material lining the room surfaces, or hanging from the ceiling.

9.2 Separation of Quieter and Noisier Machines and Activities

In general, the strategy should be, wherever possible, segregating noisy and quiet areas and trying to minimise the number of people working in the noisy areas. Ideally, noisy machines and activities should be in a separate space with the workplace (i.e., separated from quieter areas by a complete sound-insulating partition wall, with a high-sound reduction index).

9.3 Increasing Distance

According to a simple theory, the level of direct sound will reduce with the distance from the source at a rate of 6 dB per doubling of distance. In practice it may less, perhaps 4 to 5 dB per doubling distance. Therefore, if employee workstations have a direct line of sight of noisy machines, the separation between workstations and noisy areas should be made as large as possible.

9.4 Use of Partitions, Screens or Barriers

A barrier between the source and receiver blocks direct sound but there is the possibility of some sound diffusing or bending around the edges of the screen. Because the amount of sound which bends around the edges of the screen depends on the sound wavelength, the screen will be more effective for high frequencies than for low frequencies. The screen will be more effective if placed close to the source or receiver and if it is lined with sound-absorbing material on the side facing the noise source.

9.5 Using Sound-Absorbing Material

Using sound-absorbing materials, for example, slabs of mineral fibre (contained in thin polymer bags) hanging from the workplace ceiling, will reduce the amount of reverberant noise reaching workstations. This treatment will be more effective at medium and high frequencies (above the 250 Hz octave band) but will usually be limited to noise-level reductions of less than 10 dB, often to 5 or 6 dB, but this can still be very useful in some circumstances and can be very effective subjectively.

Note: use of sound absorption will reduce only reverberant sound and not the sound which travels directly from the source to the receiver. Hence, usually it will not reduce the sound that an operator receives from the machine that he/she is operating. The robustness and durability of the absorbers and their possible effect on natural lighting must also be considered.

9.6 Enclosures and Refuges

Using enclosures around noisy machinery will reduce the level of both the direct and reverberant sound produced by the enclosed machine, and be of benefit to all employees, but they will still receive noise from the other machines. A 'noise haven' (or 'noise refuge') enclosing one (or more) employee will reduce the noise from all sources, but will benefit only those employees inside the refuge.

Figures 9.1–9.3 illustrate the use of sound-absorbing materials and of an enclosure.

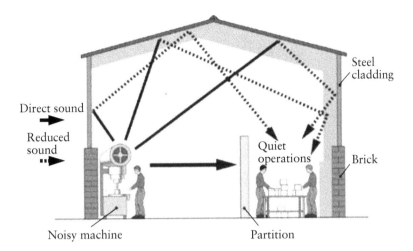

Figure 9.1 Noise paths found in a workplace. The quiet area is subjected to reflected noise from a machine somewhere else in the building. Reproduced with permission from Controlling Noise at Work L108 under *the Control of Noise at Work Regulations 2005*, Health and Safety Executive, Bootle, Merseyside, UK, 2005. ©2005, Health and Safety Executive

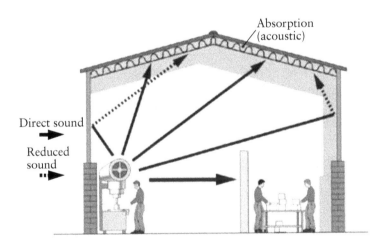

Figure 9.2 The correct use of absorption in the roof will reduce the reflected noise reaching the quiet area. Reproduced with permission from Controlling Noise at Work L108 under *the Control of Noise at Work Regulations 2005*, Health and Safety Executive, Bootle, Merseyside, UK, 2005. ©2005, Health and Safety Executive

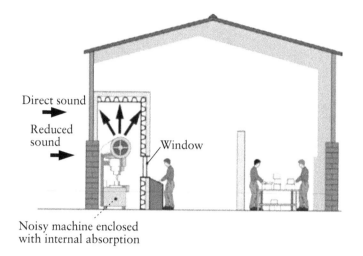

Figure 9.3 Segregation of the noisy operation will benefit the whole workplace. Reproduced with permission from Controlling Noise at Work L108 under *the Control of Noise at Work Regulations 2005*, Health and Safety Executive, Bootle, Merseyside, UK, 2005. ©2005, Health and Safety Executive

9.7 Automation

Use of automated or semi-automated processes enables equipment to be monitored and controlled remotely. This strategy ensures that employees' presence close to noisy machines and noisy areas is minimised, and so their noise-exposure levels are reduced.

Even fairly simple measures such as automated feed of material into a noisy machine ensure that the employee does not have to stand close to what is often the noisiest part of the machine. This approach can be effective in reducing the noise exposure level of the employee. Sometimes, similar treatment for the exit or delivery of material will also be effective.

9.8 Reduction of Structure-borne Sound

All machines which might transmit vibration to the building structure should be fitted with anti-vibration mounts to minimise transmission of structure-borne noise to other parts of the building.

9.9 Reduction of Noise Radiated to the External Environment

There is a need also to consider façade sound insulation of the workplace to minimise the radiation of noise to nearby noise-sensitive premises, particularly *via* apertures such as doors, windows and ventilation openings. Doors and windows facing noise-sensitive properties should have good sound insulation and be well sealed, and ventilation apertures should be fitted with acoustic attenuators. Delivery and loading areas should be sited away from noise-sensitive properties, and not have a direct line sight of them.

9.10 Location of External Machinery and Equipment

Noisy equipment located outside the workplace, including heating, ventilation and air conditioning equipment on the roof, should be located away from noise-sensitive properties.

9.11 Summary

Quiet and noisy areas should be separated by location to different parts of the building wherever possible.

Direct sound from machines may be reduced by:

- Increasing the distance between the source (the noisy machine) and the receiver;
- Interposing a screen or partition between the source and receiver;
- An enclosure around the machine; and
- An acoustic refuge for the employee.

The level of reflected (or reverberant) sound may be reduced by sound-absorbing material lining the room surfaces, or hanging from the ceiling. Noisy machine should be isolated from the building structure to minimise structure-borne sound transmission.

Use of automated or semi-automated processes will enable equipment to be monitored and controlled remotely, ensuring that the employees' presence close to noisy machines and in noisy areas is minimised.

Noise radiated to the external environment should be reduced by taking care to minimise sound transmission *via* the building façade, particularly *via* doors, windows, and ventilation openings, and by suitable location of external machinery and equipment away from noise-sensitive properties.

Bibliography

1. BS EN ISO 11690:1997 Acoustics – Recommended practice for the design of low-noise workplaces containing machinery – Part 1: Noise control strategies.

2. BS EN ISO 11690:1997 Acoustics – Recommended practice for the design of low-noise workplaces containing machinery – Part 2: Noise control measures.

3. BS EN ISO 11690:1997 Acoustics – Recommended practice for the design of low-noise workplaces containing machinery – Part 3: Sound propagation and noise prediction in workrooms.

4. Controlling Noise at Work L108 under *the Control of Noise at Work Regulations 2005*, Health and Safety Executive, Bootle, Merseyside, UK, 2005.

10 Case Studies

10.1 Introduction

Noise-control methods and principles are best explained and understood by illustrations and examples in the form of case studies. Case studies relating to noise in the workplace have been used, by the Health and Safety Executive (HSE) in particular, as a motivational tool to encourage engineers and managers in industry to attempt noise-control measures in their workplaces rather than relying on earplugs and ear muffs to protect the hearing of workers in industry.

The case studies presented in the 1985 RAPRA guide *Noise in the Plastics Processing Industry: A Practical Guide to Reducing Noise from Existing Plant and Machinery* have been summarised in **Chapter 3**.

This chapter first reviews some other existing case studies and then presents some additional ones.

10.2 Review of Existing Published Case Studies

10.2.1 HSE 100 Practical Applications of Noise Reduction Methods

In 1983, the HSE produced their publication *100 Practical Applications of Noise Reduction Methods* with a simple format of one case study per page with a few sentences explaining the nature and magnitude of the noise problem, the noise level before treatment, an explanation of the noise reduction treatment and the cost, illustrated by a photograph.

The publication is now out of print but available online at *https://archive.org/details/op1277236-1001*.

The costs are out-of-date (1983 prices), but otherwise the examples of the application of noise control are still of interest and relevance – they illustrate what can be done.

The case studies come from various industries: engineering (24), construction and transport (3), plastics (3), food and drink (5), woodworking (19), wire (4), textiles (4), printing (4), paper and board (6), concrete (11), grinding (1), materials handling (10), agriculture (1) and general – compressors and boilers (4). The principles and methods demonstrated in many of the examples could be applied to other situations and other industries.

The three plastics industry case studies were later used in the 1985 RAPRA guide *Noise in the Plastics Processing Industry: A Practical Guide to Reducing Noise from Existing Plant and Machinery*, summarised in **Chapter 3**:

• Plastics granulator treatment, including use of sound-absorbent material and baffles in the throat of the machine, cladding of machine surfaces with mastic damping compound, resulting in a noise reduction of 11 dBA, from 101 to 90 dBA.

• Substantial enclosure for a plastics granulating machine, use of a brick enclosure with careful design of a feed chute to provide 30 dBA noise reduction at the operator-feed position.

• Acoustic hopper for mobile plastics granulator, using modified feed hoppers made from double-skin 16-gauge steel with a 25 mm cavity-filled with acoustic tile. This achieved a noise reduction of 12 dBA, down to 83 dBA.

About 40 of the case studies featured acoustic enclosures, but other noise-control methods illustrated included use of: alternative processes; low-noise design; acoustic screens; panels and partitions; damping and cushioning; sound absorbers and 'acoustic refuges'. One case study illustrated the use of active noise control (ANC) (active control of gas turbine compressor noise).

Three cases involved use of ceiling absorbers fitted to workplace ceilings produced noise reductions of between 4 and 6 dBA.

10.2.2 HSE Sound Solutions Case Studies

In 1995 a second set of 60 case studies were issued by HSE in the publication *Sound Solutions*, now out of print but available online at *http://www.hse.gov.uk/Noise/casestudies/soundsolutions/index.htm*.

Many of the case studies were from the engineering, food and drinks industries, but six case studies were general in nature (i.e., not related to any particular industry) and several others could be applied to other industries. There were nine case studies featuring the use of enclosures, five illustrating reduction of noise from air jets as well as examples relating to use of damping, isolation, lagging impact noise, and use of sound absorption. There were five cases studies about noise from presses. One of these demonstrated a reduction of 4 dB in reverberant noise in a press shop as a result of hanging sound absorbers from the ceiling.

There were three case studies illustrating active noise control:

- Drinks canning plant – to reduce noise from a six-bladed radial fan in the filler room which supplied air to move cans along the lines and which produced 300 Hz pure tones.

- Inside cars – for the suppression of engine-induced 'boom'.

- Introduction of a hybrid active silencer (active absorptive) reduced discharge noise by 42 dB from a rotary blower.

10.2.3 HSE Industry Specific-Case Studies

There were three case studies specifically related to the plastics industry:

- Reducing jet noise from a plastic mould cleaning gun.
- Reducing noise from strand pelletisers by quieter design.
- Reducing noise from extrusion line cut-off saws.

The text only of two of these (the photos were not available) given in **Sections 10.2.3.1** and **10.2.3.2**.

Quieter by Design – Strand Pelletisers

The problem

Strand pelletisers are used in the plastics industry to convert continuous strands of plastic into small pellets for use in subsequent manufacturing processes. The most important part of a pelletiser is its cutting head, where the plastic strands are fed by rollers into the path of a high-speed, multi-bladed rotating head.

A-weighted noise levels from a pelletiser can exceed 90 dB. Noise is generated by the impact of each blade against the strands and the alternate compression and expansion of air as the moving knives pass the fixed bed knife edge.

The solution

A company manufacturing strand pelletisers has developed a series of new machines which incorporate several important noise-reducing features.

On examining the cutting operation, it was revealed that the speed of rotation could be reduced by increasing the number of blades fixed into the cutting head. It was found that a helical blade, which would pass progressively across the bed knife edge cutting one strand at a time rather than all simultaneously, would further reduce noise levels. Additional benefits of the helical cutter were found to be reduced wear and a reduced need to sharpen blades.

The machine casing was also redesigned, reducing both mechanical and airborne noise radiation. Anti-vibration treatment was adopted, with the cutting head installed on a base isolated from the rest of the machine.

The cost

About £25,000 for a new pelletiser and about £11,000 to upgrade a conventional pelletiser (1995).

The result

A noise reduction of more than 10 dB.

Source

Designed and manufactured by John Brown Plastics Machinery Ltd (from the HSE document Sound Solutions). Reproduced with permission from Sound Solutions, Health and Safety Executive, Bootle, Merseyside, UK, 1995. ©1995, Health and Safety Executive

Reducing Noise from Extrusion Line Cut-off Saws

The problem

Although the process of extruding plastic sections does not normally generate significant noise levels, terminal cut-off saws operating for a few seconds every 5 min can emit very high noise levels.

One company had up to 20 such saws operating in an open production area with employees on the lines being regularly exposed to A-weighted levels of 100 dB when all the lines were working.

The saws themselves were adequately protected by safety guards. However, because these were all in open steel mesh, they offered no attenuation of sound radiation from the saw blades.

The solution

The sound radiation was effectively impeded by replacing the mesh guards with solid panels, lined on the saw side with acoustically absorptive material. These were fitted both above and below the saw bench. An acoustic strip curtain was hung along the product out-feed, thereby achieving further attenuation.

The method has since been extended to cut-off saws on new lines incorporating an acoustically lined hood to swing into place over a solid base.

The cost

About £650 per hood (1995).

The result

A noise reduction of 15 dB without having either to alter the machine itself or prevent access to the saw.

Source

Courtesy of Caradon Duraplus Ltd (from the HSE document Sound Solutions). Reproduced with permission from Sound Solutions, Health and Safety Executive, Bootle, Merseyside, UK, 1995. ©1995, Health and Safety Executive

10.2.4 Environmental Noise-Related Case Studies/Histories

In 1978, the Building Research Establishment (BRE) published 23 case histories illustrating solutions to environmental noise problems in *The Control of Noise from Fixed Premises – Some Case Histories*, BRE CP 36/78. A major objective of this publication was to provide examples of noise-control measures which had been used to assist local authorities deal with similar issues using the then recently newly issued Control of Pollution Act, 1974 (Section 3 on noise pollution).

Although the costs of noise control are no longer valid, in all other ways this series of case histories remain relevant and illustrate many important features of the noise-control process.

All but two of the case histories described were associated with complaints of noise from local residents living close to industrial or commercial premises.

Noise-control methods

The case histories feature three types of noise-control measures:

- Reduction of noise at source.

- The use of standard noise-control solutions such as silencers, enclosures and screens.

- Modifications to the building structure.

Reduction in noise at the source may be achieved by replacing the existing source with a quieter piece of equipment or by modifying the method of operation, or by using damping material. In some situations it may be possible to re-site the offending noise source, and all of these approaches are illustrated in these case histories.

Several of the case histories demonstrate that several different measures may be needed to completely solve the noise problem.

Silencers

The case histories featured two types of silencers: dissipative (in which the acoustic performance depends mainly on the presence of sound-absorbing material) and reactive (in which the acoustic performance depends mainly on the geometrical shape). The former usually gives noise reductions over a relatively wide frequency range whereas the latter may be designed to achieve a considerable degree of noise reduction over a narrow band of frequencies, usually at low frequencies.

External noise sources

For about half of the examples, the noise sources which gave rise to complaints were situated outside rather than inside the main factory building, sometimes in lightweight sheds and outhouses. Examples of such sources are refrigeration and cooling equipment; dust-extraction equipment and compressors; fans and fan motors; compressors; blowers and pumps; cooling towers.

Solutions to these problems included improving sound insulation, use of enclosures and screens, use of silencers and replacement of equipment with quieter alternatives. In many cases, the problem could have been avoided by location of the plant away from noise-sensitive properties.

Internal noise sources

The next biggest class of examples involved equipment located inside the premises. In most cases, most of the noise emissions were from 'weak links' in the sound insulation of the building such as doors and windows (especially when left open in summer), and ventilation apertures, rather than from the main building fabric.

Many noise problems first come to light on warm summer nights when both factory and houses have their windows open, thus increasing the level of factory noise inside the house by less than 20 dB. If the factory windows are to be kept closed, alternative means of ventilation will need to be provided.

In particular, doors which have to be opened frequently will often be left open permanently. The problem can be reduced by ensuring that such doors do not face noise-sensitive buildings. Similarly, openings required for ventilation, be they roof ventilators, windows or ventilation ducts, should where possible be sited so that they do not face a noise-sensitive building.

Solutions to these problems include: upgrading sound insulation of the building, relocation of exhaust apertures away from noise-sensitive properties, and noise control at the source (damping, vibration isolation).

Solutions often involved several different measures, including use of silencers, damping, improved maintenance, and redirection of exhaust extract/outlets.

Boilers and boiler houses

Five of the case histories related to noise from boilers and boiler houses which involve combustion noise from burners, exhaust fans, and exhaust stacks, which can give rise to environmental noise problems.

Assessment of environmental noise – tonality and 'excess over background'

Unlike occupational noise problems in many of these cases (about one-third), the problem was of low-frequency tonal noise, often in the evenings and night-time, and frequency band analysis rather than measurement of dBA levels was required. In many cases, the problems were solved by measures which reduced the tonal content of the noise, even though the overall dBA level had not been reduced significantly.

In other cases, where there was no tonal content there was evidence that complaints usually arose when the noise being complained of clearly exceeded the ambient noise level in the absence of the offending noise – which is very broadly the principle underlying the assessment method used in BS 4142:2014.

Three of these case histories are given in **Sections 10.2.4.1.1–.10.2.4.1.3.**

BRE CP 36/78 (Uttley and Heppell)

Three case histories from *The Control of Noise from Fixed Premises – Some Case Histories*, BRE Current Paper 36/78.

BRE Case History: Case 7

Noise at night from a plastics factory adjacent to a residential area.

Summary

Noise at night from a plastics factory adjacent to a residential area gave rise to complaints. The noise levels were reduced by increasing the sound insulation of the building which housed the offending machinery.

Noise sources and location

The factory was situated on the edge of a trading estate which adjoined a residential area (**Figure 10.1**). Significant noise sources consisted of compressors, blowers and cooling pumps which were housed in a lightweight structure attached to the main building and facing a row of houses. A shredder and reciprocating compressor in the main building were other possible sources of noise.

Noise measurements and remedial action

Noise measurements were made outside 3 houses and showed levels of 52, 56 and 50 dBA with the factory in operation and a night-time ambient level of 36 dBA.

The lightweight structure which houses the machinery outside the main building was replaced by a brick structure lined with sound-absorbent material. The noise levels outside the 3 houses were re-measured and found to be 47, 48 and 44 dBA, respectively.

Although still above the ambient, these levels were found to be acceptable to the local residents so no further action was taken.

Further comments

A modulated pure tone which was originally thought to come from the plastics factory was eventually traced to an adjacent factory. It may be that this pure tone originally sparked-off the complaints.

Further action which was proposed (but not taken) included the fitting of silencers to the cooling fans and stopping using the shredder at night.

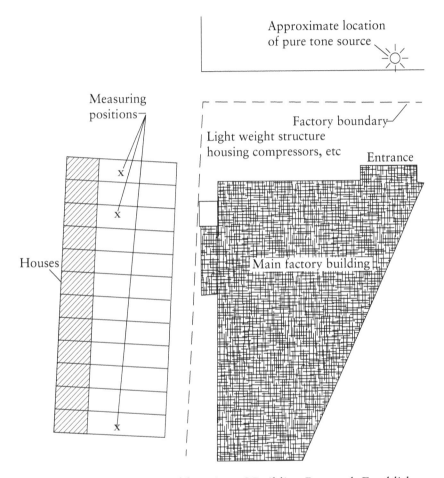

Figure 10.1 Noise sources and location. ©Building Research Establishment

BRE Case History: Case 14

High-pitched noise produced by bandsaws which caused complaints at night.

Summary

A high-pitched noise produced by bandsaws caused complaints at night. The noise level was reduced at the source by damping the drive wheels of the bandsaws.

Noise sources and location

Two bandsaws were situated in a small corrugated asbestos shed detached from the main factory building (**Figure 10.2**). The bandsaws cut tubular mild steel which was used for the main work of the factory. The shed had no windows but large sliding doors.

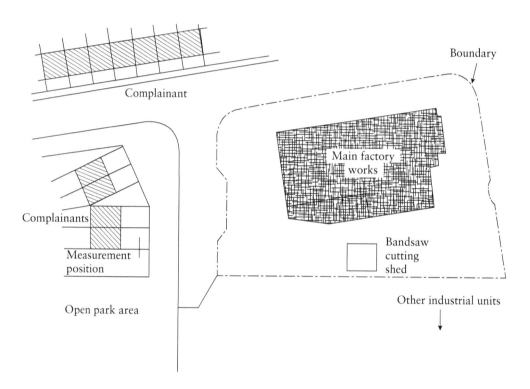

Figure 10.2 Location of bandsaws. ©Building Research Establishment

Noise measurements and remedial action

Octave band noise measurements were taken outside the nearest complainants' house with the bandsaws in use, without the bandsaws in use but with the rest of the factory working normally, and with the factory closed. In addition, noise levels were measured inside the shed with the bandsaws in operation. These levels together with the levels measured at the nearest complainants' house after treatment had been undertaken (**Table 10.1**).

Table 10.1 Bandsaw noise measurements							
	Octave band centre frequency (Hz)						dBA
	125	250	500	1,000	2,000	4,000	
Ambient with factory closed	43	37	34	36	29	24	39
Factory operating without bandsaws	54	52	46	40	31	29	47
Factory operating with bandsaws	54	46	46	42	47	28	51
Inside shed with bandsaws in use	87	83	84	82	90	76	94
After treatment	49	43	38	33	28	24	44
©Building Research Establishment							

Initial proposals were for screening of the bandsaws and rebuilding of the shed, but it was decided that this would be too expensive. The noise levels were eventually reduced by fixing rubber damping rings to the main drive wheels of the bandsaw. This treatment effectively damped the resonance at about 2,000 Hz and resulted in a reduction of almost 20 dB in the noise level at that frequency. Simultaneously, new doors were fitted to the shed which housed the bandsaws and this reduced the general noise that occurred at other frequencies.

Further comments

The tonal quality made the noise which the bandsaws produced particularly noticeable and hence unacceptable to the local residents.

By reducing the noise at source, the high levels inside the shed were also controlled.

The cost of treatment was probably less than the original proposal for rebuilding the shed to increase its sound insulation.

BRE Case History: Case 17

Tonal noise emitted from the stacks of a boiler house that was causing complaints from residents nearby.

Summary

Tonal noise was being emitted from the stacks of a boiler house that was situated in the centre of a large telecommunications establishment. The noise from the boilers was reduced at the source by modifications to burners and by replacing the primary air fan.

Noise source and location

This establishment was part of a large telecommunications centre. The boiler house was surrounded by other buildings and was not visible from the complainants' houses. The complainants lived ≈50 m from the boiler house, as shown in **Figure 10.3**.

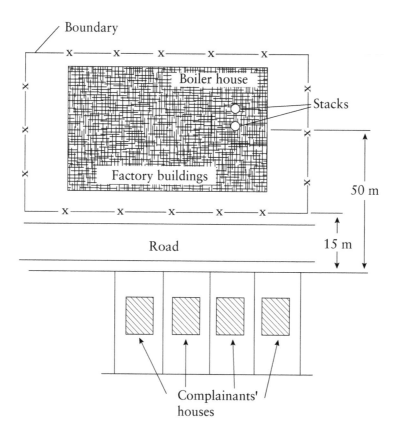

Figure 10.3 Complainants' properties in relation to boiler house.
©Building Research Establishment

Following some modifications to the boiler house, high levels of tonal noise were emanating from the stacks. The length of the stacks corresponded to two wavelengths at 84 Hz. It was concluded that the stacks were resonating because of tonal noise produced from within the boilers.

Examination of the equipment inside the boiler house revealed two possible causes of the 84 Hz tone: the atomising cups in the burner units or the dampers on the primary air fan intakes.

The atomising cup was designed to run at 5,200 rpm but it was believed to be running at a slightly slower speed. If the shedding of fuel by the cup was not uniform throughout the rotation cycle, then fluctuations in combustion could give rise to noise with a strong tonal component at 86 Hz. This was believed to be the primary cause.

The rating of the primary air fan was considerably in excess of that required so consequently the intake damper was virtually closed. The frequency spectrum of the noise produced by the passage of air between the damper blades was calculated to peak at 83 Hz.

Noise measurements and remedial action

Overall A-weighted and octave band noise levels were measured on several occasions at the boundary fence and inside the complainants' house. The levels measured at the boundary fence are given in **Table 10.2**.

Table 10.2 Octave band sound levels relating to the investigation of tonal noise emitted from the stacks of a boiler house								
	Octave band centre frequency (Hz)							dBA
	63	125	25	500	1,000	2,000	4,000	
Unmodified boiler	62	53	42	36	33	27	20	43
Modified boiler	55	49	40	35	33	27	20	39
Background	52	46	40	34	33	26	19	38
©Building Research Establishment								

It was proposed to alter temporarily the rotational speed of the atomising cup to demonstrate whether the fan or the cup was the primary source. However, the client changed the burner unit atomising cup rotational speed at the same time as installing a smaller-capacity fan. The noise produced at the boundary fence by the modified boiler was reduced by 4 and 7 dBA in the octave band of interest.

Further comments

The weather at the times of the tests was dry with fresh winds. Recordings were made, therefore, during lulls in the wind.

As a result of these alterations there was a slight reduction in boiler efficiency.

Environment Agency Case Studies Relating to Environmental Noise

The Environment Agency has published detailed reports of six case studies to illustrate various aspects of noise control in the environment. Collectively they illustrate a wide range of noise-reduction measures and strategies. The individual case studies are summarised below but the reader should refer to the detailed reports themselves to appreciate the care and detailed preparation that is needed to achieve success (*Best Available Techniques for Control of Noise & Vibration*, R&D Technical Report P4-079/TR/1, 2001).

Noise from a Foundry

Noise reduction from the foundry was based around the introduction of a new type of furnace system (a change from coke-fired to electric furnace) installed into a purpose-built building, which was also designed to contain the scrap-handling facility and to minimise noise emissions to the environment. The foundry operated for 24 h each working day.

The change was planned to minimise noise to the environment, and to meet noise limits using the advice of noise consultants with detailed noise surveys carried out before, during and after the installation.

Measured noise levels from the original installation were up to 20 dB above background noise in nearby residential areas and low-frequency tonal noise, and bursts of impulse noise from materials handling were clearly audible. The main sources of noise were from the furnace and materials handling, including scrap metal, and from fume-extraction equipment.

The change to the electric furnace meant that the wet scrubber, which had been a major source of noise in the old system, was no longer necessary and was removed. The old dry-extraction systems were replaced with new low-noise systems. A new dust-extraction plant was installed for the electric melt facility and was located to achieve shielding of noise to nearby noise-sensitive areas by the new building.

The new building was constructed on a steel portal frame with dense concrete blockwork to a height of 2 m and double-skinned composite profiled metal cladding above to the eaves. The roof of the building was also constructed of double-skinned composite metal cladding.

The main access door was designed to be large enough to allow an articulated lorry with a tipping trailer to exit with the trailer raised. This presented technical problems due to the effects of wind on such a large area, and for sealing the door when closed. The solution was to fit a top-hung door with flexible rubber strip seals all around the perimeter.

The result of these changes was to reduce overall noise levels by 12 dB and achieved a substantial noise improvement for the local community.

The new arrangements produced several additional environmental, operational and financial benefits as well as noise reductions.

Noise from a grain-malting facility in a rural environment

This case study describes the noise-reduction programme carried out at a large maltings operating in a rural area. Some of the plant operated for 24 h a day. Residents living at the boundary of the site had complained of the noise, and the local authority was carrying out an investigation into possible noise nuisance.

The main noise sources were several continuously operating fans which produced broadband noise with tonal components, grain conveyor systems which produced bursts of impulsive noise (creaks, groans, clatter, screeches and clonks) and grain chutes which produced high-frequency 'hissing' sounds.

Noise surveys had shown that the continuous noise from the fans was well above the background noise at the site boundary, which in itself was likely to cause annoyance, and was exacerbated by the tonal quality of the fan noise and the additional noise form the conveyors and chutes.

Two noisy centrifugal blower fans were used in the grain-drying process, and in addition fans were used to operate the dust-extraction system at the main grain input from delivery lorries, and a third system of fans operated an aeration plant for the fermentation process.

Fan noise was reduced using a series of in-line absorptive silencers. Some fans also required acoustic lagging to reduce noise breaking out from the fan casings. Noise reductions of about 5 dBA were achieved and tonal noise was eliminated.

Conveyor and elevator casings were fitted with dense polyvinyl chloride matting to provide acoustic damping. A grain chute treated in this way showed a noise reduction of more than 10 dBA. Noise from conveyor and elevator drive mechanisms were boxed in with small local enclosures manufactured on site from sheet steel and mineral fibre board.

An overall noise reduction of 7 dBA was achieved and though overall noise levels remained above the background noise, they were reduced to below recognised absolute noise standards and the noise from the conveyor and chutes was reduced to acceptable levels.

This case study illustrates how the impact of environmental noise was reduced using a series of in-house, low cost, low-tech solutions.

Products factory

The factory produced various types of paper products from dry paper pulp which was transported around the factory through 600 mm-diameter ducts. This air flow transport system was powered by a series of fans which produced continuous pure tone noise at a frequency of about 160 Hz and intermittent bursts of high-frequency noise when air was released to atmosphere at a baler exhaust.

Although the factory was located in a rural environment there were two dwellings within 500 m with a direct line of sight of the plant. These neighbours had complained about the noise to the local authority. A BS 4142:2014 assessment indicated that reductions to both types of noise were necessary.

The tonal noise was reduced using four Helmhotz resonator-type silencers. These consisted of the introduction of side-branch ducts 1 m long attached at right angles to the main duct transport system. An adjustable piston was fitted inside the resonator operated by a screw thread in the closed end to 'tune' the resonator to the correct frequency. Although the overall noise level from the fans was reduced by only 2 dB, the tone was eliminated. The noise from the baler-exhaust noise was also reduced to inaudible levels by the use of partial enclosure, which allowed unimpaired air release around each exhaust outlet.

An interesting feature of this case study was the careful consideration and eventual rejection of several different alternative noise-control solutions.

A major safety factor affecting choice of the noise-control solution was the risk of fire ignition and dust explosion in the duct system. This ruled out the use of conventional absorptive exhaust silencers and in-line duct absorptive or reactive silencers which might cause dust accumulation. Other possible solutions that were eliminated for reasons of cost effectiveness or other reasons were the reduction of fan speed, replacement by quieter fans, and active noise control.

Mineral fibre manufacturing plant

Complaints of a low-frequency tonal noise during the night from residential premises situated about 100 m from a plant making mineral fibre insulation products led to the local authority declaring that a statutory nuisance existed.

The noise was only audible within the dwelling at certain times depending on the levels of background noise and was most noticeable in the complainants' bedroom at night-time. It was of a very low frequency and accompanied by a distinctive 'beat'.

After careful investigation the noise was identified as coming from two vibrating screens used in the production process. It could be measured outside the dwelling, even though it could not be heard, using 1/3 octave bands analysis, and shown to lie within the 25 Hz 1/3 octave band.

The screens were made to vibrate by an eccentric weight which rotated at low speed and produced an out-of-balance force which caused the vibration.

The effect of reducing the operating speed of the vibrator and of only using one screen was investigated by correlating levels of the tone measured close to the screen with those measured both inside and outside the dwelling, and comparing noise levels with the screen switched on and off.

A reduced operating speed using one screen only (which eliminated the beats caused by the two screens operating together) was found. This satisfied the requirements of the production process and reduced the level of the tone by 10 dB, which eliminated the problem.

Noise from a gas turbine-combined heat and power generator at a food production plant

A coal-fired boiler system which supplies steam to a food factory was to be replaced by a combined heat and power (CHP) unit to reduce pollution as well as to save energy and running costs.

The new plant, which was to be located 200 m from the nearest residential premises, was subject to strict local authority planning conditions: the existing background noise level of 40 dBA and a Noise Rating of 35 should not be exceeded at the nearest dwelling.

The proposed new system will comprise two gas turbines and associated heat-recovery boilers and gas compressors, and an ancillary plant including three gas-fired boilers, a water-treatment plant and a standby diesel generator to ensure a back-up supply of electricity in the event of a failure.

Several of these items of plant will each incorporate multiple sources of noise such as from air inlet and exhausts, combustion exhaust, cooling fans and noise radiating directly from plant surfaces.

The strategy chosen to comply with the noise limits was to enclose all items of the plant within one new building, rather than to control each sound source separately. With this arrangement, noise emissions from the plant to the external environment would arise from three sources:

- Noise from all the various air inlets and outlets feeding the various plant items and including the ventilation requirement of the building.

- Sound radiated directly from (i.e., breaking out from) the building structure.

- Noise from the single exhaust stack carrying combustion exhausts from all the various items of the plant.

To protect the hearing of personnel working inside the CHP building, it was also necessary that internal noise levels should be reduced to less than 85 dBA. This required separate enclosures inside the building for the two turbines and the standby generator, and acoustic lagging to ductwork.

This case study describes in detail the process of designing the facility to meet the planning noise limits, involving the specification of sound insulation (sound reduction index values) of the building fabric and insertion loss values for attenuators, in octave bands. Specialised stainless-steel hot-gas attenuators, with purpose-made infills manufactured from basaltic mineral fibre, were required to deal with the high temperature and pressures in the gas turbines.

Breakout from the building structure was limited to acceptable levels through the use of 140 mm dense (2,000 kg/m^3) blockwork in the building wall construction. The roof of the building was constructed from composite metal cladding materials comprising: 0.7 mm outer steel panel; 200 mm cavity with 80 mm glass fibre quilt infill at a density of 12 kg/m^3; 0.7 mm steel inner-liner panel.

The gas turbine-bypass exhausts, the heat-recovery boiler exhausts and the standby diesel generator were designed to vent through a single stack. To reduce regenerated noise in the stack, a limiting gas velocity of 20 m/s was specified.

High levels of noise inside the building arising from breakout from the ductwork between the turbines exhausts and the stacks were reduced by lagging the ductwork in an envelope of mineral wool (density of 128 kg/m^3) clad with aluminium sheet to a thickness of 0.5 mm.

On commissioning of the plant, noise surveys at the nearest residential dwellings demonstrated that the design and implementation had been successful and that there had been no increase in ambient noise.

Control of noise from landfill site operations by low-cost best practicable means

A proposed landfill operation in a rural area close to two noise-sensitive properties was subject to noise-level limits set by the local planning authority.

Modelling of possible future noise levels using BS 5228:2009 methodologies suggested that substantial noise control was required to meet the noise limits.

This case study describes in detail the noise-control measures devised to meet the noise limits: construction of two earth bunds to act as noise barriers to shield the noise-sensitive properties, and a series of operational measures.

As a result of a survey of background noise and a thorough review of all noise sources and noise operations likely to be involved, a package of noise-management practices was developed: the maintenance and switching off of equipment when not in use; limiting sound power levels in the plant; restricted operating times; limited use of audible warning alarms; noise monitoring.

The overall noise reduction achieved was about 10 dBA and, although levels were still above background noise levels, they met the requirements of the local authority.

Overall summary

These case studies illustrated the use of detailed noise surveys to: establish noise levels close to items at a plant and at the boundaries to noise-sensitive premises; identify the contributions of individual noise impact to the overall impact; use of BS 4142:2014 as the main noise assessment tool; response to nuisance complaints and local-authority involvement.

10.3 Additional Case Studies

This chapter concludes with six additional case studies. Five of these relate to the plastics industry and one relates to noise from a boiler, which could arise in almost any industry. Four of them relate particularly to noise in the workplace and two to noise in the environment.

The first case study, from D. Bull, dB Acoustics, Colchester, UK, was first published in *Noise Control (A Pira Environmental Guide)* in 2000.

The second case study describes an active noise control system of a plastic resin manufacturing plant in the USA, first installed in 1987 but still in operation.

This is followed by a description of an investigation into noise from a boiler.

The next two case studies describe the approach to noise control by two manufacturers using plastic moulding processes.

The sequence ends with a series of examples of noise control provided by a leading noise-control consultancy specialising in noise control at source and in fan-noise reduction.

Noise Control in the Production of Sheet Plastic Film

The purpose of this case study is to demonstrate what can be achieved on an in-house basis by an engineer who has been properly trained in noise control applying more-or-less standard principles and techniques in a systematic and persistent manner.

The process starts with plastic granules which are heated, injection moulded and extruded to produce sheets of plastic film. Reject material is recycled through a granulator.

The main noise sources are many different fans, the granulators, jet noise from discharges of cooling air, and noise radiated from large-diameter pipes conveying granules.

The case study concerns a factory where the average overall factory noise was reduced from 100 to less than 85 dBA. Apart from in a small number of areas nobody on the factory floor now has to wear ear protectors. **Figure 10.4** shows the layout of one part of the factory and lists ten measures that were introduced to achieve this noise reduction.

The reduction was achieved by one of the company's own engineers who had attended a training course in the measurement and control of noise. The starting point was careful measurement of existing noise levels in all parts of the factory. A noise map, giving dBA contours, resulting from such measurements for another part of the factory is shown in **Figure 10.5**. This was used to establish priorities for noise-control action, on a step-by-step basis, which in this case included enclosures around noisy machines, a noise haven, replacement of old, noisy equipment by a quieter plant, in-line silencers to air blowers, and the fitting of better-quality gears to some machines.

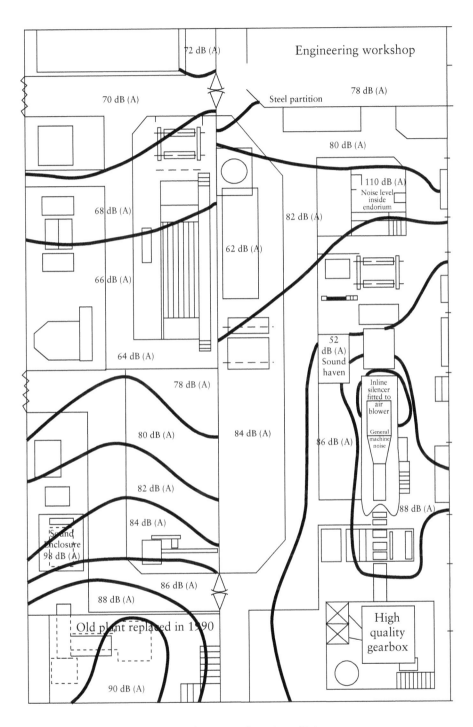

Figure 10.4 Noise map showing dBA contours.
©dB Acoustics

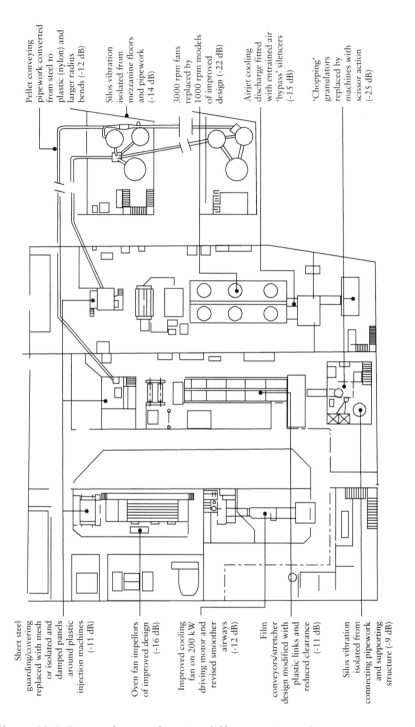

Figure 10.5 Factory layout showing different noise-reduction measures.
©dB Acoustics

Acknowledgement

This case study has been provided by D. Bull of dB Acoustics, 25 Albany Road, West Bergholt, Colchester, UK.

An Industrial Application of Active Noise Control

The first installation, dating to 1987, was performed on a collection of around 2 induced-draft fans located on the roof of a plastic resin manufacturing plant in Wisconsin, USA.

The speakers and microphones are mounted on existing discharge stacks. Cables from those components are run in conduit to controllers located indoors nearby. Mechanical and electrical work amounts to approximately 1 day of labour for each unit. However, unlike the weight burden of passive silencers, no heavy-lift equipment, nor stack supports, are required. Thus, the installed cost is favourable compared with the alternatives.

The speakers are located on the side wall of the stacks so as to not restrict the airflow, allowing the fans to operate at the same rpm, airflow and static pressure load as originally designed.

The fans draw air from inside the plant through fabric filters. As such, the temperature is moderate at approximately 80 °F (27 °C), and the exhaust flow is clean, so the speakers and microphones are in a 'benign' environment. Typical operation at the plant is day-time only, whereupon the speakers and microphones cool to winter temperatures below freezing at night, but have no problem with functioning properly the following mornings. Limiting loudspeakers to one-quarter of maximum excursion, lifetimes can exceed 15 years in clean environments. In applications where speaker excursions approach the maximum, shorter lifetimes have resulted. The fabric filters typically require bag replacement at least yearly, so the cost and timing of occasional speaker replacement is a negligible fraction of the operating expense for the fan system in general.

Generally, there is redundancy in speaker selection such that active noise control function continues to be achieved even with one or more speakers in need of replacement.

Acoustical performance

Commonly used backward-curved centrifugal fans mostly had problem noise at the blade-pass frequency. These low-frequency tones were dramatically eliminated by the ANC system. Owing to various combinations of the number of blades and fan speeds

229

for the different installations, the resulting composite noise spectrum measured at a central roof location, seen in **Figure 10.6**, showed a series of pure tones dominating a lower-broadband level. Individual tones were reduced by more than 30 dB on a narrowband basis, and less than 20 dB in an octave band analysis. **Figure 10.7** shows a typical installation where four loudspeakers for the active noise control system are mounted in the fan exhaust.

Overall dBA reductions could be modest, usually less than 10 dB, in part because the reduction occurred at low frequencies. However, change as small as 3 dBA could still have a dramatic subjective impact. In fact, many ordinances had 5–10 dB penalties for tonal spectra, so elimination of the problem tones could have been a bonus even in objective on/off evaluations.

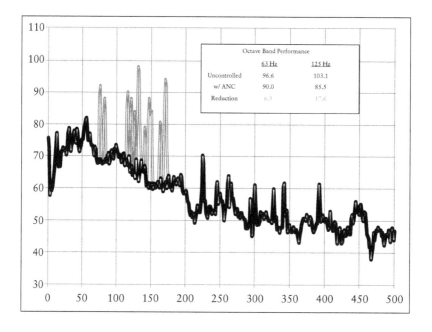

Figure 10.6 Combined fan-exhaust noise spectra with and without active noise control. ©Wise Associates

Figure 10.7 Typical installation (fan is shown blue at the bottom of the figure). Four loudspeakers for the active noise control system are mounted in the fan exhaust. Reproduced with permission from Wise Associates. ©Wise Associates

This was an analysis of the noise from all the operating systems which had different fan frequencies and levels.

Acknowledgement

This case study has been provided by Steve Wise of Wise Associates, 1409 East Skyline Drive, Madison, WI, 53705, USA.

Control of Low-frequency Noise from a Boiler

A resident complained of a low-frequency 'rumble' that he believed to be coming from a factory 120 m away. A single logging noise meter was put in the complainants' unoccupied bedroom, which had a grazing line of sight to the factory. Levels here showed regular 5–10 dB exceedances of the 100 Hz set by the Department for Environment Food & Rural Affairs NANR45 criteria [1] as illustrated in **Figure 10.8**. Subjectively, the audio recordings sounded like a bulk air handling unit rather than a mains hum (a common source of 100 Hz noise), and Fast Fourier-Transform analysis found the frequency to be 95–108 Hz.

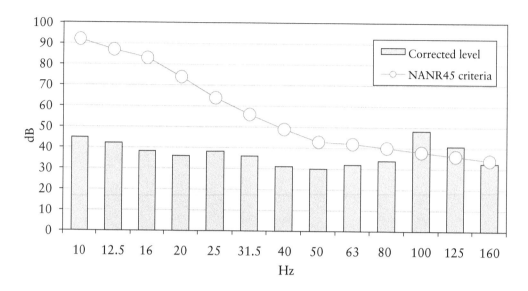

Figure 10.8 Low-frequency 1/3 octave levels *versus* NANR45 criteria.
©Environment Agency

The site processes were switched off sequentially, which clearly indicated that the noise source was the site's large boilers. These boilers had recently been changed from being oil-fired to gas-fired to reduce air pollution. The boilers were later changed back temporarily to oil, and the low-frequency noise levels were reduced.

The reason for the greater noise from burning gas instead of oil was possibly due to the natural gas (predominantly methane) having a greater flame velocity than the slow burn rate of the larger hydrocarbons found in kerosene. This greater energy could then be 'exciting' one of the acoustic modes of the boiler furnace. It is very difficult to predict the frequency of this mode because there will be dynamic changes and gradients in gas density, velocity and temperature within the furnace. Thermoacoustic vibrations, where the furnace behaves like a Rijke or Sondhauss tube [2], can be ruled out as the temperature gradients from burning oil and gas are essentially the same.

The flame plate in the boiler where the flame is 'held' was moved further away from the fuel jets and some of the jets were closed off to improve the fuel-air mixing prior to ignition (**Figure 10.9**). This resulted in a 3 dB reduction at the boiler, which was likely due to improved flame stability.

Further attenuation was required, and the operator chose to install a large-absorptive silencer (3.8 m long and 0.5 m wide, with an inner annular ring) (**Figure 10.10**), which resulted in a dynamic insertion loss of 25 dB at approximately 100 Hz, as shown in **Figure 10.11.** Following these modifications, no further complaints were received

Figure 10.9 Flame plate. ©Environment Agency

Figure 10.10 Installed absorptive silencer. ©Environment Agency

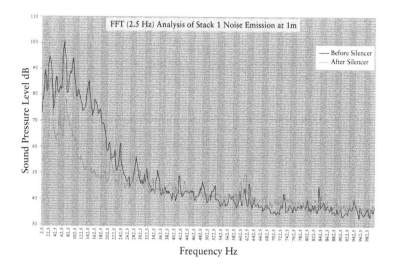

Figure 10.11 Dynamic insertion loss of the silencer. ©Environment Agency

Acknowledgement

This case study was provided by J. Tofts of the Environment Agency, Rio House, Waterside Drive, Aztec West, Almondsbury, Bristol, UK.

Plastic Moulding Machines – The Stewart Company, Banbury

The Stewart Company is a UK market leader in the manufacture and supply of technical plastic moulded products for the gardening, home and professional catering sectors. The company offers a large range of products in its portfolio, ranging from garden planters and propagators, to food containers and specialist storage products. The company employs up to 120 people at its manufacturing, warehousing and distribution centre in Banbury, UK.

The products are created using 18 injection-moulding machines and 3 rotational moulding machines. Raw material in the form of plastic pellets is delivered to the site and stored in silos. These are transported to the moulding machines by a vacuum pipework system powered by four compressors. Scrap plastic material is recycled using a grinding machine.

A plan of the ground floor of the site is shown in **Figure 10.12**. The production area is divided into four bays. Bay 1 contains the rotational moulding area and storage racks, bay 2 contains the injection-moulding area, and bays 3 and 4 are devoted to storage, packaging and despatch. There are also designated hearing-protection zones where ear protectors must be worn at all times. These are enclosed areas which comprise the pump/compressor room, the tool room and grinder room.

The person responsible for noise levels at the site is M. Hewer, IMS Coordinator, and the health and safety committee. The company is certified to BS EN ISO 9001:2008.

Although there is no formal 'buy quiet' policy in place, noise levels will always be a consideration when replacing machinery.

The company commissions a workplace noise assessment approximately every 2 years by an external independent occupational hygiene company, unless changes in machinery or working practice require a more frequent update.

The company also employs the services of a commercial mobile audiometric testing company to carry out regular health surveillance of those employees whose personal daily noise-exposure level are shown by the noise-exposure assessments to be above the first action level of 80 dBA.

Noise-exposure Assessment Report

The latest workplace noise survey consisted of a combination of dose badge tests and a series of short-duration sample noise level measurements. Noise-dose badges were fitted to 10 employees for a 4 h period. These employees were selected on the basis of working in the noisiest areas and, therefore, expected to have the highest noise-dose values. The data collected by each dose badge gave a measurement of the overall noise-dose level over the 4 h period, and also a minute-by-minute history showing the minute-by-minute variation of the noise level throughout the 4 h period. Two examples of these time histories are shown in **Figures 10.13** and **10.14**. The dose badge readings were confirmed and supplemented by short-duration sample noise level measurements taken with a sound-level meter at 63 locations round the workplace (**Figure 10.12**). The dose badge tests covered the following areas: one from the grinding room, three from the rotomoulding areas (bay 1), two from maintenance (including in bay 2) and 3 from operatives working in bay 2.

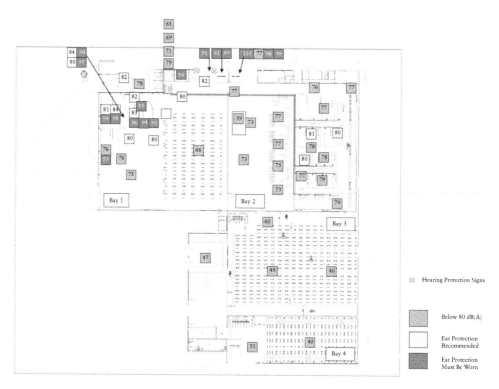

Figure 10.12 Map showing noise-measurement locations and noise levels (A-weighted equivalent sound level, LAeq). ©Stewart Company

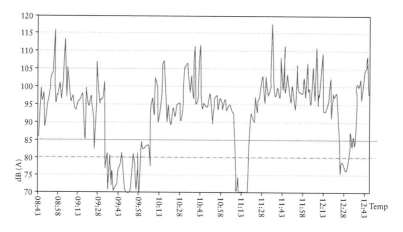

Figure 10.13 Example of noise-dose time history (1 min intervals) – grinding room. Reproduced with permission from The Stewart Group. ©Stewart Group

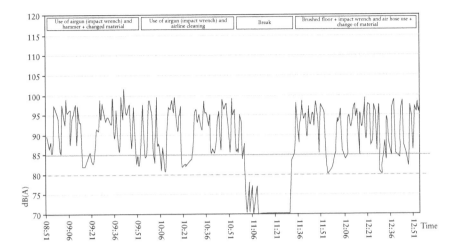

Figure 10.14 Example of noise-dose time history (1 min intervals) – rotary moulder operator. Reproduced with permission from The Stewart Group. ©Stewart Group

In addition, octave-band noise level measurements were carried out for locations where the noise levels were more than 85 dBA so that the effectiveness of hearing protection could be evaluated. Two types of ear protectors were in use: a lightweight 'behind the head' semi-aural canal cap type of earplug, and a heavy duty-type of earmuff. These calculations included a 4 dB reduction of attenuation to allow for difference of performance in use as compared with that under standard test conditions, as recommended by HSE guidance. The results of calculations were presented and indicated the sound level at the ear with the ear protectors being used and fitted correctly, and a check that this level was below the required level of 80 dBA.

Each employee fitted with a dose badge was asked to complete a questionnaire to establish where they were working and which activities they were performing during the 4 h period of the noise-exposure monitoring exercise. This information, together with the data from the dose badges and the noise measurements, were used to produce estimates of the daily personal noise exposure (LEp,d) of the 10 selected employees. These estimated noise-exposure level values were presented in the report and compared with the noise-exposure action values of the *Control of Noise at Work Regulations 2005* and indicated, for each of the 10 employees, whether their noise exposure fell into one of three categories relating to the requirements of the *Control of Noise at Work Regulations 2005*:

- Below the lower-exposure action value of 80 dBA.

- Above the lower-exposure action value, but below the upper-exposure action value of 85 dBA.

- Above the upper-exposure action value of 85 dBA.

The report outlined the duties of employers under the regulations for each of these three categories of noise exposure.

Noise levels

The highest of the short-turn sample noise level measurements occurred in the hearing-protection zones:

- Pump/compressor room 94 dBA.

- Grinder room 99 dBA.

- Tool room use of stippler tool (110 dBA), use of a mallet (for banging moulds to eject product, 98 dBA), use of air gun impact wrench (99 dBA).

Although the noise emitted from the rotary extruders in bay 1 was about 80 dB, much higher noise-exposure levels were experienced by employees working in the mould bays using mallets (for 'knocking out' products from the moulds) an air gun (impact wrench) or in the finishing areas as a result of using air lines (for cleaning products prior to painting), a stipple gun, or pneumatic drill (for drilling holes in plant pots).

In bay 2, the noise emission from the injection moulders was also about 80 dBA at their product ejection outlets, but a system of conveyors transported the products away from the injection moulders to storage packing and despatch areas where the noise levels experienced by employees involved in these tasks were several dB lower.

Occasional maintenance and setting up work which involved operatives working close to the outlet feed hoppers would mean exposure to the slightly higher noise levels, of about 80 dBA. The sound levels at bays 3 and 4 were well below 80 dBA

Peak action level

The sound-level meter also recorded the C-weighted peak sound level (LCpeak) at each noise-measurement position. Peak action levels of the regulations were not exceeded at any of the 63 noise-monitoring positions.

Noise to the external environment

Noise levels up to 82 dBA were measured outside the building from fan exhaust points and at chiller units and extraction units. Noise emission to the environment was not a problem at this site because of its location on an industrial site well away from noise-sensitive properties.

Recommendations

The report gave recommendations to reduce noise-exposure levels, including:

- Investigate use of alternative or quieter tools and processes, for example, alternatives to hammering for removing products from moulds;

- Fitting noise-reducing jet nozzles to airlines for cleaning, or reduce air pressure or use alternative cleaning methods, such as vacuum cleaners, or advising operatives to use ear protectors when using existing airlines for cleaning; extending hearing protection zones; and

- Advising operatives to use ear protectors when using air lines and the stippler, considering alternative quieter alternatives for an air wrench; use of heavy-duty earmuffs only (i.e. not canal cap-type earplugs) when using the impact wrench and stippler; consider introduction of a 'buy quiet' policy for new equipment.

Changes under consideration

In response to these recommendations, the company is in the process of reviewing and considering various possible changes, involving re-location of new and existing equipment and aimed at reducing noise-exposure levels in certain areas. These include:

- Use of differing types of mould-release methods to remove the need for 'knocking out'; the use of vacuum cleaners for cleaning;

- Installation of an isolated blast booth off-line for all tools to be cleaned on removal;

- Quieter alternatives for air wrenches;

- Installation of a new rotational moulding machine, with inclusion of sound-reducing baffles;

- Changes to the layout of hearing-protection zones is in progress; and

- The issue and use of hearing protection is being reviewed.

These changes will require a new noise survey upon completion.

Acknowledgement

This case study has been provided by Mr. M. Hewer of Stewart Company, Banbury, OX16 1RH, UK.

Noise from Plastic Moulding Processes

This case study has been contributed by a large plastics processing group of companies. The group is a leader in worldwide design and engineering of plastic products and unique in supplying products made by all processes: blow moulding, injection moulding, rotomoulding, flexibles and thermoforming. The group operates from over 120 production sites in over 30 countries, ranging in size from about 50 to almost 1,000 employees, of which typically about 70% are employed in production areas, where noise levels can be high and requiring hearing protection to be worn.

The information for this case study has been provided as a result of a meeting with the senior member of the group responsible for coordination of health and safety within the group. The case study reflects practice related to noise and hearing protection throughout the companies in the group.

Noise and hearing conservation policy

Noise and hearing conservation are dealt with as part as a comprehensive health and safety policy, rather than as a separate issue.

Implementation is the responsibility of the site general manager and with the full involvement of the health and safety committee.

Noise-making processes and equipment

Noise-making processes and equipment are: granulators, injection moulders, blow moulders, injection stretch blow moulders, rotomoulders, thermoformers, extruders, hydraulic systems (ancillary), and compressed air systems – mainly on downstream equipment.

Noise-exposure assessments

Noise-exposure assessments are carried out at least every 18 months, or more frequently when changes are made to the production area, such as the introduction of new machinery, as well as changes in layout and production practices which could affect noise-exposure levels of employees. These assessments are carried out by external specialist companies, though many sites have equipment on site to carry out *ad hoc* tests.

Action plans and noise-reduction measures

Action plans are drawn up following noise-exposure assessments. There is a continuous effort to reduce noise levels where possible. There is no 'one size fits all' approach, but with actions depending on the site, equipment, environment and operation.

Examples of noise-reduction measures which have been introduced include:

• Acoustic damping of noise-radiating surfaces – enclosures where necessary.

• Acoustic curtains – overlapping but removable to ensure ease of maintenance and replacement – checked as part of the shift safety checks.

• External ducting of exhausts.

• Silencers and diffusers on air systems.

• Acoustic lagging of pipework.

Pneumatic and hydraulic equipment is being replaced by quieter and less energy-consuming electrically operated equipment, where practical, featuring the use of servo and stepper motors, solenoids as opposed to hydraulic or pneumatically operated actuators.

Motor-gearbox systems are being changed to inverter drives on air transport systems to reduce noise levels (by operating at controlled lower speed). There is a perpetual review of technology in this area.

Use of hearing protection

It is company policy that the wearing of hearing protectors is compulsory at all times in production areas. This policy aims to ensure protection is maximised and that employees have a safe working environment. It also avoids the problem of employees having to remove or re-fit hearing protection as they pass from noisier to quieter parts of production areas, simplifying the rules for all.

Employees are given a choice of protectors (earplugs or earmuffs) although advice may be given for the most suitable types in certain areas. Many sites provide custom-made hearing protection for full-time staff.

Warning signs are displayed at entry points to areas where protectors must be worn. In some areas, in outside yards, for example, where the wearing of hearing protection could give rise to a hazards from moving vehicle, warning signs are displayed to remove ear protectors.

Fork lift truck drivers may be issued with earmuffs fitted with active noise control cancellation so that they do not have to remove or re-fit protectors as they drive in and out of production areas. Trials are ongoing with earmuffs fitted with bandpass/frequency filters covering the speech frequency range, which can be switched in or out to help them communicate with other people while in a hearing-protection zone without removing their protectors.

New employees are given a combination of face-to-face and online training about the use of hearing protectors as part of their induction programmes, supplemented as necessary as part of their regular programme of periodic hearing tests. Hearing protection is issued during induction of new employees and signed for by both the recipient and issuer.

The wearing of hearing protectors [together with other personal protective equipment (PPE)] is regularly monitored. Monthly checks on the use of all PPE are carried out and signed by the shift manager. In addition there are inspections by shift supervisors and safety managers, and regular health and safety inspection tours by members of the health and safety committee. These inspections can include the use of photographs to record examples both of good practice and compliance.

Health surveillance

The company uses external organisations to carry out audiometric hearing tests on all employees working in production areas as well as during their pre-employment induction process.

Buy quiet policy

Noise emission is always a significant (but not an overriding) factor which is considered when purchasing new equipment unless the equipment is to be used outside and may be subject to complaints from neighbours. It might sometimes be considered preferable to make in-house noise-reduction measures on a newly purchased item of equipment (e.g., granulator) when installed on site using the company's own engineers rather than purchase more expensive so-called 'quiet' products to ensure that both the equipment itself and its operating environment are taken into account when ensuring that the noise emitted is minimised. The quality of noise-emission data provided by manufacturers of equipment tends to be unsatisfactory.

Record-Keeping

The company keeps records relating to several aspects relating to noise control and hearing conservation matters:

- Reports of noise surveys/risk assessments.

- Introduction of noise-control measures. Where appropriate, these are broadcast throughout the group to spread good practice.

- Issue of ear protectors.

- Training given to employees.

- Health surveillance records.

The keeping of such records is important for defending the company against hearing-loss claims from former employees, as well as providing evidence of compliance with the *Control of Noise at Work Regulations 2005* in the UK, and corresponding legislation in other countries.

Hearing loss/industrial deafness claims

As with many similar companies, the group receives claims for compensation for industrial deafness from former employees. It is, generally, understood that the frequency of such claims throughout British industry has increased rapidly in the last few years as a result of the introduction of a 'no win no fee' claims culture (e.g., articles on the Association of British Insurers website at *http://abi.org.uk*).

Although the vast majority of such claims are spurious, significant cost (including solicitors' legal fees) and time are spent in defending them. This is one reason why such emphasis is placed within the company on thorough record-keeping (including photographic evidence), as indicated above.

Other recommendations to minimise resources spent on defending against such claims include: collecting statements from employees stating how noise is treated and use of PPE within the company; carrying out pre-employment hearing checks and, if possible, pre-retirement checks as well.

Environmental noise

Most of the group's production sites are located in industrial estates so complaints about environmental-noise problems are not common.

Such problems occur mainly at sites where land surrounding the factory has been sold and developed for residential accommodation. In such sites a 'buy quiet' policy is more likely to be carefully employed and additional care is taken to purchase a rooftop extract and ventilation plant with the lowest possible noise emission values.

One example of the successful resolution of an environmental-noise problem illustrates the importance of systematic and persistent diagnosis in solving noise problems, as indicated below.

The complaint was of noise late at night from a factory in an urban area. Noise surveys at ground level in the surrounding area late at night had failed to detect the noise or its source. The situation was finally diagnosed by a visit by the investigating engineer to the sole complainant, who lived at the top of a high-rise residential block, with a direct line of sight of the roof of the company's factory.

From this elevated position, the source of the noise was identified as the impact of polymer pellets striking the walls of a duct and being transmitted to the atmosphere *via* the extract duct system.

Once identified, the problem was readily solved using standard noise-control techniques: lagging the duct walls and erection of a parapet wall around the plant to acoustically shield the reception position from direct line of sight of the noise source.

10.4 Case Studies Provided by the Industrial Noise & Vibration Centre

These case studies are based upon extracts from the Industrial Noise & Vibration Centre (INVC) presentation *Noise Management and Control 2007 Plastics Industry* in 2007, and on extracts from noise-control guides published by INVC.

INVC Approach to Noise Control

Noise control is not a safety issue – noise control is an engineering problem that should be solved by engineering means, in particular through noise control at source.

Best practice in noise control can only be achieved by following a simple diagnostic process that provides an engineering insight:

* Effective noise control must be based on an accurate diagnosis and not on assumptions.

* All the options must be considered, not just the conventional high-cost palliatives of enclosures and silencers. These techniques should be used only if it can be proved that there is no engineering alternative.

Noise Control Audit

The noise control audit is an alternative to carrying out costly placebo risk assessments that tell you what you already know: you have a noise problem. It provides the information you need to determine the options to eliminate noise problems using techniques that can be self-financing surprisingly often.

Objectives:

* Assess company noise-control options using the best technology.

* Generate cost *versus* noise reduction trade-offs for each item of a noisy plant.

- Plan a practical and cost-effective noise-control programme.

The results of the audit also take into account factors such as:

- Hygiene: access/maintenance

- Productivity

Surprisingly often, implementing good engineering noise-control measures will improve productivity and reduce costs.

If the audit proves that control is impractical, it also provides certification so that PPE can be used for long-term risk control.

'Best practicable' means noise control audit steps:

- List all the potential noise sources on each piece of the noisy plant.

- Rank the sources.

- Assess all the noise-control options for the dominant source:

    - Potential reduction in noise from this source

    - Operational, productivity, hygiene and operator constraints

    - Cost

- If engineering control is not practical for the dominant source, then you have proved that screening/enclosure are the only options.

The results are used to generate cost *versus* noise reduction trade-offs for each item of a noisy plant and to plan the most practical and cost-effective noise-control programme possible across the company.

The INVC Quiet Fan Technology Approach to Reducing Noise from Fans

INVC have taken a similar approach to that of the way that Formula 1 teams invest in the design of aerodynamic aids to control the airflow round their cars. Fan noise is the sum of the turbulence-generated pressure fluctuations in the air shed by the blades, so INVC have developed a range of aerodynamic inserts that are fitted inside the fan casing to smooth the air flow. This reduces the pressure fluctuations – and hence the noise – at the source without introducing the back pressure often associated with silencers. This not only reduces the noise travelling down the intake and exhaust ductwork, but also that passing through the fan casing which may not only eliminate the need for silencers, but also the need for acoustic enclosures.

Some quiet-fan technology solutions are featured in several of the following case studies.

1: Extruder drive noise control

Extruder drive noise decreased by about 10 dBA *via* fan tone/cooling silencing and guard damping (**Figure 10.15**).

Precise analysis of sources: 93 dBA – silence fan tone; 85 dBA implemented quiet-fan technology (aerodynamic noise control insert fitted inside fan casing) to remove tone – 8 dBA attenuation (**Figure 10.15**).

Drive motor modifications (damp guard + silence cooling system): approximately 82 dBA lamination of thin metal guards to provide very efficient damping that reduced vibration radiated as noise. Simple acoustic-absorbent lined shroud over motor cooling vents – additional 3 dB reduction.

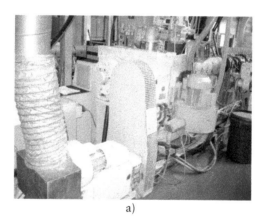

a)

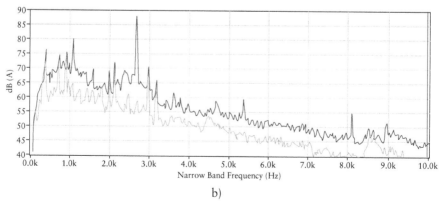

b)

Figure 10.15 Control of noise from an extruder drive

2: Noise control from an extruder vacuum system (**Figures 10.16**)

Noise levels were reduced from 91 dBA to less than 85 dBA by the use of a manifold silencer. There are several air exhaust outlets. These were manifolded by running a single pipe along the unit into which each exhaust was plumbed. A single silencer (rear exhaust box from a car) was then fitted to the end of the pipe.

a)

b)

Figure 10.16 Control of noise from an extruder vacuum system

3: Control of granulator noise (**Figures 10.17** and **10.18**)

Granulator (and similar plant) noise decreased by less than 18 dBA by fitting low-cost web dampers in place of an enclosure – 100 dBA sound level varying over time dominated by web-radiated noise – 18 dBA reduction *via* web damper on a similar machine.

On many web-fed granulators, the noise is dominated by web-vibration radiated as noise. In this example, the proof that this was the dominant component was provided by the drop in noise level when the web was perforated (to reduce the noise-radiating surface area). Damping rollers were fitted to the web near the outlet to prevent the transmission of vibration into the large web area to provide 18 dB attenuation.

Figure 10.17 A perforated web to reduce noise

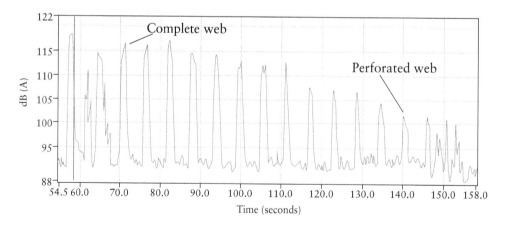

Figure 10.18 Noise reductions due to perforated web

4: Pellet and component transport system (**Figures 10.19**)

The thin walls of the pipework-radiated pellet impacts as noise. The walls were damped by gluing and clamping formed sheet steel to create an *in situ* laminate. In the pellet transport system, the noise level was typically dBA at 1 m, whereas pipe dampers (**Figure 10.19**) gave a 10–20 dB reduction.

a)

b)

Figure 10.19 Noise control in a pellet and component transport systems. (a) Pipe dampers and (b) noise-reduced plastic granule air transport fan (one side removed)

Fan tones were eliminated by fitting internal aerodynamic inserts to reduce the noise at the source. Key noise-radiating thin sheet steel panels were also damped (laminate) and the duct geometry changed to make the system more efficient. In the component transport fan system (95–100 dBA), quiet-fan technology gave a 10 dBA reduction, and overall the noise decreased to less than 85 dBA.

5: Ultrasonic plastic welding (**Figure 10.20**)

Ultrasonic plastic welding (99 dBA) was dominated by 8 and 16 kHz octave bands. Analysis of an emailed video showed that the dominant noise was a very-high-frequency airborne sound. This was highly directional, so a simple small clear shield (with a small area of high-frequency sound-absorbent material on other surfaces to reduce reflections) provided a substantial noise reduction – acoustic close shields reduced the noise level to less than 85 dBA.

Figure 10.20 Close shield to reduce noise in ultrasonic plastic welding

6: Filler cooler pipes (**Figures 10.21**)

Problem: 94 dBA from cooling pipes for sealed toothpaste tube ends – rapid cooling a necessity.

Conventional solution: enclosure – high cost with hygiene and productivity issues. The proposed enclosure would have been very expensive and would also have caused production problems because access was needed to tubes that were sometimes knocked out of the holders by the airflow. The coanda nozzles (in place of the perforated copper tubes) provided more efficient cooling, used 20% less air and the reduced turbulence meant that tubes were no longer knocked from the holders, thereby improving productivity.

Best solution:

- Coanda effect linear nozzles – 12 dBA noise reduction (82 dBA)

- Improved performance (less turbulence-disturbing tubes)

- 20% less air consumption – pay for themselves very quickly

- No effect on access or operation

a)

b)

Figure 10.21 Noise-reduced filler cooler pipes

7: Hopper vibrators (**Figure 10.22**)

Vibratory feed hopper noise-control measures (damping and geometry changes) doubled throughput, eliminated down-time caused by fatigue cracking, and reduced power consumption by 80% and noise by 22 dBA:

- Noise-radiating surfaces were laminated to provide very high damping.

- Sometimes, simple geometry changes are all that is required to improve flow and reduce the requirement for vibration. Vessels and chutes should also be isolated so that only the component in contact with the materials being fed is vibrated.

- Damping surfaces (lamination) reduces high-frequency vibration but not low-frequency feed vibration.

- A vibrator was connected to the internal mesh fitted inside the frame through a sleeved hole in the hopper wall. This resulted in higher levels of vibration being transmitted into the material, but very little vibration in the hopper.

- Where there was 'hammer rash' (impacts to the chute or hopper to free stuck material), the dents were repaired (because these cause more flow problems) and, if necessary, an 'anvil' was fitted so that future impacts did not dent.

Summary:

- Design angles of chute.

- Vibration isolation of chute to reduce the required vibrator amplitude.

- Isolate the vibrator from the chute – tuned to amplify low frequency and attenuates high frequencies.

- Damp chute to reduce high frequencies.

- Vibrate material inside rather than in the whole chute.

- Fit purpose-designed anvil to allow hammering without damage.

- Noise reduction: 22 dBA.

Figure 10.22 Noise reduction from hopper vibrators

8: Vibratory bottle unscrambler (**Figures 10.23**)

Problem: vibratory feed hopper for unscrambler generating 90 dBA.

Best solution:

- Poor design: vibrators at high level to vibrate whole hopper – feed still inefficient.

  A laminated plate was fitted inside the existing hopper (using grommets for isolation), connected to the vibrators through holes in the hopper: cost about £400.

- Reduced vibrator level, improved feed, noise reduced from 90 down to 82 dBA.

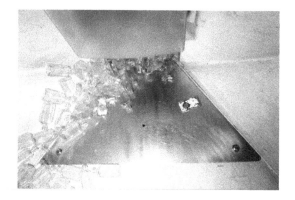

a)

b)

c)

Figure 10.23 Noise reduction for vibratory bottle unscrambler

9: Scrap fan extract and chopper fans (**Figures 10.24** and **10.25**)

Scrap fan noise reduction – both safety and environmental-noise issues arising from tonal noise.

Fan silencing without silencers, attenuators, lagging or acoustic enclosures. Three sets of chopper fans were used to pull-off scrap cans from the lines and to shred them for recycling. These generated high levels of tonal noise, causing both occupational- and environmental-noise problems. Instead of fitting conventional fan attenuators, acoustic enclosures and noise lagging at a potential cost of more than £30,000, the INVC quiet-fan technology was fitted to the fans in a matter of hours to give an overall noise reduction of 22 dBA at an installation cost of about £1,000/line – a capital cost saving of 90% and with no effect on fan performance or efficiency.

Unlike fan silencers, the modifications were unaffected by the passage of scrap cans and will last the lifetime of the fans without maintenance – and less than 1% of the previous noise level.

Figure 10.24 Scrap fan extract and chopper fans

Figure 10.25 Noise-reduction measures for scrap fan extract and chopper fans

10: Verso press used in the manufacture of electrical components (**Figure 10.26**)

Problem: high-speed strip fed press at 101 dBA in a quiet area.

Conventional solution: manufacturer-supplied 'acoustic guards' gave only 3 dBA reduction. Full enclosure suggested.

BPM solution: noise control audit showed dominant source to be fabricated press legs.

Tico vibration isolation pads fitted between frame and legs tuned to give a natural frequency of 65 Hz for both legs (different loads) – 9 dBA noise reduction at the closest operator position – cost £45 for materials and 1 man-day to fit.

Figure 10.26 Vibration isolation of Verso press

11: Weighing machine enclosures (**Figure 10.27**)

Problem: typically 87–98 dBA – high hygiene.

Convention enclosure: enclosures – about 5 dBA reduction, and often increases operator noise exposure by 2–3 dBA – cost £8,000–15,000 + capital + access/hygiene/maintenance problems.

For best engineering control INVC uses engineering noise-control measures instead of enclosures (damping, geometry, cushioning). Typically this results in 20% of the cost of enclosures and provides more attenuation (up to 20 dBA depending on the product). This also improves productivity because enclosures are usually no longer necessary – +10 dBA reduction, 20% of the cost – ×4 performance maintenance, access and cleaning as for unmodified machine.

Summary:

- 94 dBA with enclosure – 82 dBA without enclosure; PPE unnecessary; and improved productivity, cleaning, access and maintenance.

- The noise sources in the weighing machine are: product impacts on chutes and hoppers, plus hopper mechanical impacts.

- Geometry changes and constrained layer damping are the best solutions.

Figure 10.27 Weighing machine enclosure

12: Quiet tape (**Figure 10.28**)

Quiet tape uses stronger glue – generates more tension; may have to adjust machines to use the new tape.

Tape is available with an adhesive pattern and formulation that reduces the noise from carton taping operations – these can generate more than 100 dBA. Typical noise reductions can be over 20 dBA – it may be necessary to change the tape tension.

Figure 10.28 Quiet tape

13: Pneumatic silencers and nozzles (**Figures 10.29–10.31**)

Silencer solutions:

- Zero back-pressure silencers
- Standardise on three sizes
- Fix piped silencers to machine and manifold multiple exhausts

Air-entraining nozzles:

- About 10 dB quieter for the same thrust
- Use about 20% less air
- Pay for themselves very quickly
- Intrinsically 'safe'

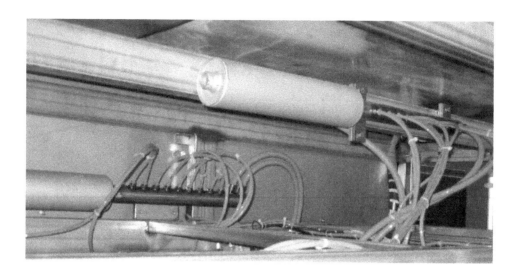

Figure 10.29 Fix piped silencers to machine and manifold multiple exhausts

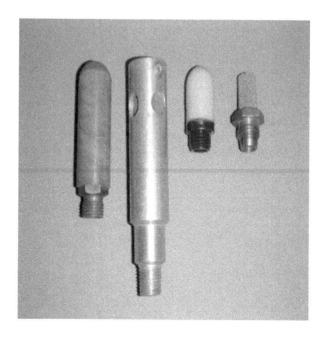

Figure 10.30 Types of pneumatic silencers

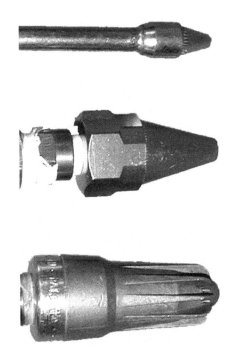

Figure 10.31 Types of air-entraining nozzles

Bibliography

1. *Noise in the Plastics Processing Industry: A Practical Guide to Reducing Noise from Existing Plant and Machinery*, RAPRA, Shawbury, UK, 1985.

2. *100 Practical Applications of Noise Reduction Methods*, Health and Safety Executive, Bootle, Merseyside, UK, 1983. *https://archive.org/details/op1277236-1001*

3. *Sound Solutions*, Health and Safety Executive, Bootle, Merseyside, UK, 1995. *http://www.hse.gov.uk/Noise/casestudies/soundsolutions/index.htm*

4. W.A. Utley and T.W. Heppell in *The Control of Noise from Fixed Premises – Some Case Histories*, BRE Current Paper 36/78, Building Research Establishment (BRE), Watford, UK, 1978.

5. Control of Pollution Act, 1974.

6. BS 4142:2014 – Methods for rating and assessing industrial and commercial sound.

7. S. Mitchell in *Best Available Techniques for Control of Noise & Vibration*, R&D TECHNICAL REPORT P4-079/TR/1, Environment Agency, Bristol, UK, 2001.

8. BS 5228:2009 – Code of practice for noise and vibration control on construction and open sites.

9. *Noise Control (A Pira Environmental Guide)*, Pira, Leatherhead, UK, 2000.

10. A. Moorhouse, D. Waddington and M. Adams in *Proposed Criteria for the Assessment of Low Frequency Noise Disturbance*, Department for Environment, Food and Rural Affairs, London, 2005.

11. F.L. Eisingser and R.E. Sullivan, *Journal of Pressure Vessel Technology*, 2002, **124**, 418.

12. BS EN ISO 9001:2008 – Quality management systems – Requirements.

13. *Buy Quiet*, Health and Safety Executive, Bootle, Merseyside, UK, 2010.

14. *Control of Noise at Work Regulations 2005*, Guidance on the Regulations, Health and Safety Executive, Bootle, Merseyside, UK, 2005.

15. *Noise Management and Control 2007 Plastics Industry*, Industrial Noise & Vibration Centre (INVC), Slough, UK, 2007.
   http://www.invc.co.uk/invc/pdfs/invc-noise-control-plastics-07.pdf

Glossary

1. Terms Relating to the Properties and Measurement of Sound

A-weighted decibels – A frequency weighting devised to attempt to take into account the fact that the human response to sound is not equally sensitive to all frequencies; it consists of an electronic filter in a sound-level meter, which attempts to build this variability into the indicated noise-level reading so that it will correlate, approximately, with the human response (defined in BS EN ISO 61672-1:2013).

Broadband sound/noise – Sound or noise containing a wide range of frequencies.

C-weighting – One of the frequency weightings defined in BS EN ISO 61672-1:2013; it corresponds to the 100-Phon contour and is the closest to the linear or unweighted value. Sound or noise levels measured using the C weighting are called 'C- weighted decibels', also written as dBC.

Continuous equivalent noise level of a time-varying noise (LAeq,T); This is the steady noise level (usually in dBA) which, over the period of time (T), under consideration, contains the same amount of (A-weighted) sound energy as the time-varying noise, over the same period of time. It is also called the time averaged sound level' and 'average sound level.

Decibels, the decibel scale – The decibel (dB) scale is a logarithmic scale for comparing the ratios of two powers, or of quantities related to power, such as sound pressures and sound power levels.

One of the reasons for using the dB scale is to compress this enormous range of sound pressures (from 20 micro-Pa to about 100 Pa) to a more manageable range, from about 0 to 120 dB.

Fast- (F) and Slow (S)-time weightings – An averaging time used in sound-level meters, and defined in BS EN ISO 61672-1:2013. Fast- and Slow-time weightings are used when measuring instantaneous moment-to-moment variations in sound pressure [as opposed to measuring average of equivalent levels ($L_{Aeq,T}$) over a period of time].

Frequency – The number of cycles per second of a sound wave, measured in Hertz (Hz). Strictly speaking, the definition refers to the simplest type of sound wave of just one single frequency, called a pure tone, for which the sound pressure varies sinusoidally with time. Although most sounds are mixtures of several frequencies, called broadband sound. The human hearing range extends from about 20 to 20,000 Hz (20 kHz).

Frequency analysis – Sound-level measurements in which broadband sounds are measured in a range of frequency bands, most commonly in octave bands, although 1/3 octave bands are also used, particularly for measurements of building acoustics. Octave-band sound levels are used for the design and specification of noise-control measures.

Narrowband analysis is used in the diagnosis and identification of tonal noise sources such as fans and gears. Narrowband analyses may be on a constant percentage bandwidth basis, such as 1/6, 1/12 or 1/24 octaves; or as constant bandwidth such as 100 or 10 Hz. An octave band is the range between two frequencies whose ratio is 2:1. The standard set of octave bands has centre frequencies at 63, 125, 250, 500, 1,000, 2,000, 4,000 and 8,000 Hz.

Frequency weightings – A set of electronic filters built into a sound-level meter to simulate, approximately, the average human hearing frequency response, which is most sensitive in the frequency range from 1,000 to 4,000 Hz, and less sensitive at lower and higher frequencies. Three frequency weightings, A, C and Z, are defined in BS EN ISO 61672-1:2013; see also under A- and C-weighting.

Maximum and peak sound pressure levels (L_{Amax}) – The maximum RMS A-weighted sound pressure level occurring within a specified time period; the time weighting, F or S, is usually specified. L_{Amax} values are used in determining the noise impact of short duration bursts of a high level of environmental noise.

L_{Amax} are the highest values of the sound pressure waveform of very short impact or impulsive sounds, without the application of the F or S wave forms.

Peak levels (LC_{peak}) are used in the evaluation of noise exposure in the workplace to determine compliance with peak action levels.

LC_{peak} are always higher than maximum sound levels.

Pascal (Pa) – A unit of pressure equal to 1 N/m^2 atmospheric pressure is about 100,000 Pa. An sound pressure level of 1 Pa corresponds to a sound pressure level of 94 dB.

Sound power, sound power level (L_w) – The sound energy radiated per unit time by a sound source, measured in watts (W); sound power measured on a dB scale: $L_W = 10 \log 10 \, (W/W_0)$, where W_0 is the reference value of sound power, 10^{-12} W.

Sound pressure – The fluctuations in air pressure, from the steady atmospheric pressure, created by sound, measured in Pa. The audible range of sound pressure ranges from 20 micropascals to about 100 Pa. One of the reasons for using the dB scale is to compress this enormous range of sound pressures to a more manageable range, from about 1 to 120 dB.

Sound pressure level (L_p) – Sound pressure measured on a dB scale: $L_p = 20 \log 10$ (p/p_0), where p is the sound pressure in Pa and p_0 is the reference sound pressure, 20×10^{-6} Pa.

Watt (W) – The unit of power; the power dissipated when one joule of energy is expended in one second.

Wavelength – The minimum distance between two points that are in phase within a medium transmitting a progressive wave. Wavelength is inversely related to frequency, so that high frequencies have short wavelengths and low frequencies have long wavelengths. The wavelength of the sound determines how sound bends around corners and the directionality of sound sources.

Z weighting – A (zero) frequency weighting defined in BS EN ISO 61672-1:2013. Sound or noise levels measured using the Z weighting are called Z- weighted decibels, also written as dBZ.

2. Terms Relating to the Human Response to Sound

Audiometry, audiogram – The measurement of hearing; a chart or graph of hearing level against frequency.

Daily personal noise exposure (Lep,d) – That steady or constant level which, over 8 h, contains the same amount of A-weighted sound energy as is received by the subject during the working day.

Ear defenders or protectors – Earmuffs or earplugs worn to provide attenuation of sounds reaching the ear and reduce the risk of noise-induced hearing loss.

Exposure limit value – A noise-exposure level defined in the *Control of Noise at Work Regulations 2005* requiring action from employers [a personal daily (or weekly) noise-exposure level of 87 dBA or LC_{peak} of 140 dBC].

Hearing level – A measured threshold of hearing, expressed in dB relative to a specified standard threshold for normal hearing.

Hearing threshold – For a given listener, the lowest sound pressure level of a particular sound that can be heard under specified measurement conditions, assuming that the sound reaching the ears from other sources is negligible.

Lower-exposure action value – A noise-exposure level defined in the *Control of Noise at Work Regulations 2005* requiring action from employers and employees [a personal daily (or weekly) noise-exposure level of 80 dBA, or LC_{peak} level of 135 dBC].

Noise-induced hearing loss – Hearing loss arising from prolonged exposure to high levels of noise.

Permanent threshold shift – The component of threshold shift which shows no progressive reduction with the passage of time when the apparent cause has been removed.

Presbycusis – Hearing loss, mainly of high frequencies, that occurs with advancing age.

Pure tones, tonality – A pure tone is a sound for which the waveform is a sine wave (i.e., for which the sound pressure varies sinusoidally with time).

The presence of tonal characteristics in sounds (which are often described by words such as whine, hum, drone or whistle) are often considered to be more disturbing or annoying than broadband sound without such tonal components, so are given a penalty or character correction in assessment methods such as BS 4142:2014.

Temporary threshold shift – The component of threshold shift which shows progressive reduction with the passage of time when the apparent source has been removed.

Tinnitus – A subjective sense of noises in the head or ringing in the ears for which there is no observable cause. It can often be associated with noise-induced hearing loss.

Upper-exposure action value – A noise-exposure level defined in the *Control of Noise at Work Regulations 2005* requiring action from employers and employees [a personal daily (or weekly) noise-exposure level of 85 dBA, or LC_{peak} level of 137 dBC].

3. Terms Relating to the Way Sound is Generated and Transmitted from the Source to Receiver

Airborne sound – Sound or noise radiated directly from a source, such as a loudspeaker or machine, into the surrounding air (in contrast to structure-borne sound).

Background noise, background noise level – Defined in BS 4142:2014 as the value of the A-weighted residual noise at the assessment position that is exceeded for 90% of a given time interval 'T' (i.e., $L_{A90,T}$). Measured using time weighting 'F' and quoted to the nearest whole number of dB (also see under residual noise).

Barrier attenuation – The sound reduction, in decibels, which occurs at the receive position from the insertion of a screen or barrier in between the sound source and the receiver.

Diffraction – The process whereby an acoustic wave is disturbed and its energy redistributed in space as a result of an obstacle in its path. The relative size of the sound wavelength and the object are always important in diffraction. Reflection may be considered to be a special case of diffraction when the size of the obstacle is very large compared with the wavelength. The combined effects of diffraction from an irregular array of objects in the path of the sound is also known as scattering; diffraction theory deals with all aspects of the interactions between matter (i.e., obstacles) and waves, so it also determines the directional patterns of sound radiation from vibrating objects.

Direct sound field – Sound which arrives at the receiver having travelled directly from the source, without reflection.

Far field of a sound source – The part of the sound field of the source where the sound pressure and acoustic particle velocity are substantially in phase, and the sound intensity is inversely proportional to the square of the distance from the source.

Free-field conditions – A situation in which the radiation from a sound source is completely unaffected by the presence of any reflecting boundaries (see also under *anechoic*).

Ground attenuation – An attenuation of sound at a distance from a receiver caused by interference between sound waves travelling directly from the source to receiver, and sound arriving at the receiver after reflection at the ground.

Natural frequency – The frequency of free or natural vibrations of a vibrating system, caused by a transient force such as impact. Impacts can cause undamped sheet steel panels to ring at their natural frequency, causing noise to be generated in addition to that arising from the impact itself (see under *impact noise* in **Section 4**).

Near field of a sound source – The region of space surrounding the source where sound pressure and acoustic particle velocity are not in phase, and the sound pressure varies with position in a complex way.

The practical consequence of near/far field is that sound measurements taken in the far field of a sound source can be used to predict the sound level which will be produced at location further from the source.

Reflection of sound – The redirection of waves which occurs at a boundary between media when the size of the boundary interface is large compared with the wavelength.

Refraction – The change in direction of sound waves caused by changes in the sound speed in the medium. Refraction of sound in the atmosphere due to variations in the weather (wind and temperature gradients) is responsible for day-to-day variations in propagation of sound over long distances.

Resonance – The situation in which the amplitude of forced vibration of a system reaches a maximum at a certain forcing frequency (called the resonance frequency). Resonance can be caused in rotating machinery (fans, motors, gears) when the rotation frequency produces periodic forces which excite noise radiating panels on the machine to vibrate at their natural frequency and radiate noise.

Resonance frequency – The frequency at which resonance occurs (i.e., at which the forced vibration amplitude in response to a force of constant amplitude is a maximum). For an undamped system, the resonance frequency is the same as the natural frequency of the system; for a damped system the resonance frequency is slightly reduced.

Reverberant sound, reverberation – The sound in an enclosed space which results from repeated reflections at the boundaries.

Standing waves – A wave system characterised by a stationary pattern of amplitude distribution in space arising from the interference of progressive waves; also called stationary waves.

Structure-borne sound – Structure-borne sound reaches the receiver after travelling from the source *via* a building or machine structure. Structure-borne sound travels very efficiently in buildings and is more difficult to predict than airborne sound.

4. Terms Relating to the Control of Sound and to Noise Reduction

Absorption – 1) The process whereby sound energy is converted into heat, leading to a reduction in sound pressure level and 2) the property of a material which allows it to absorb sound energy.

Absorption coefficient – A measure of the effectiveness of materials as sound absorbers; it is the ratio of the sound energy absorbed or transmitted (i.e., not reflected) by a surface to the total sound energy incident upon that surface. The value of the coefficient varies from 0 (for very poor absorbers and good reflectors) to 1 (for very good absorbers and poor reflectors).

Acoustic enclosure – A structure built around a machine to reduce noise.

Acoustic lagging – Materials applied externally to the surface of pipes and ducts to reduce the radiation of noise; not to be confused with thermal lagging.

Active noise control – A noise control system which uses antiphase signals from loudspeakers to reduce noise by destructive interference.

Aerodynamic noise – Sound/noise generated in a fluid by a disturbance of the fluid causing vibration and sound radiation.

Anti-vibration mounts/vibration isolators – Springs or other resilient materials used to reduce vibration (and noise) by isolating the source from its surroundings.

Attenuation – A general term used to indicate the reduction of noise or vibration, by whatever method or for whatever reason, and the amount (usually in dB) by which it is reduced.

Cavity absorber or Helmholtz resonator – A type of sound absorber deigned to absorb sound over only a selected very narrow range of frequencies, usually in the low-frequency range. It consists of an air-filled enclosure, or cavity, connected to the open air by a narrow column or neck.

Damping – A process whereby vibrational energy is converted into heat through some frictional mechanism, thus causing the level of vibration to decrease. Damping materials are used to reduce the 'ringing' of impacting components(e.g., in bins, hoppers and conveyors) and to reduce the vibration and noise caused by resonant systems.

Impact noise – Sound resulting from the impact between colliding bodies. Impacts can generate both airborne sound and structure-borne sound. Impact noise can be reduced by reducing the momentum of the impacting components, for example, by cushioning the impact or reducing the drop height, or by damping the components to minimise the sound producing 'ringing', which continues after the impact.

Insertion loss – A measure of the effectiveness of noise-control devices such as silencers and enclosures; the insertion loss of a device is the difference, in dB, between the noise level with and without the device present. Insertion loss will usually vary with frequency, and so needs to be specified in octave bands.

Isolation – The reduction of vibration and structure-borne sound by the use of resilient materials inserted in the transmission path between source and receiver. More generally, the term 'isolation' refers to the separation of noise source from noise-sensitive receivers.

Mass law – An approximate relationship for predicting the sound-reduction index of panels and partitions, based only on the surface density of the panel and the frequency of the sound.

Panel absorber – A type of sound-absorber designed to absorb sound over only a selected range of frequencies, usually in the low-frequency range. It consists of a panel of plywood, or heavy felt, or similar material, with an air cavity behind it.

Porous absorber – A layer of open-celled foam and fibrous material which absorbs sound depending on the porosity and thickness of the layer.

Silencer, Attenuator – A device introduced into air or gas-flow systems to reduce noise. Absorptive types contain sound-absorbing materials; whereas reactive types are designed to tune out noise at particular frequencies.

Sound insulation – The reduction or attenuation of airborne sound by a solid partition between the source and receiver; this may be a building partition (e.g., a floor, wall or ceiling), a screen or barrier, or an acoustic enclosure.

Sound-reduction index – A measure of the airborne sound-insulating properties in a particular frequency band of a material in the form of a panel or partition, or of a building element such as a wall, window or floor. It is measured in dB: $R = 10 \log_{10}(1/t)$, where 't' is the sound-transmission coefficient – it is measured under laboratory conditions according to BS EN ISO 10140-1; also known as transmission loss.

Specific noise source – The noise source under investigation for assessing the likelihood of adverse comments (defined in BS 4142:2014).

Transmissibility of a vibrating system – The non-dimensional ratio of vibration amplitude at two points in a system. Frequently, the two points are on either side of springs used as anti-vibration mounts, and the transmissibility is used as an indicator of the effectiveness of the isolation.

Transmission loss – A measure of the airborne sound-insulating properties in a particular frequency band of a material in the form of a panel or partition or of a building element such as a wall, window or floor. It is measured in dB: $R = 10 \log 10 (1/t)$, where 't' is the sound-transmission coefficient – it is measured under laboratory conditions according to BS EN ISO 10140-1; also known as the sound reduction index.

Abbreviations

AC	Alternating current
ANC	Active noise control
APV	Assumed protection values
BRE	Building Research Establishment
BS	British Standards
CE	Conformité Européenne
CHP	Combined heat and power
dB	Decibel(s)
dBA	A-weighted decibel(s)
dBC	C-weighted decibel(s)
dbZ	Z-weighted decibel(s)
DC	Direct current
EN	European Norms
EU	European Union
F	Fast-time weighting
HML	High, medium, and low method
HSE	Health and Safety Executive
HVAC	Heating, ventilation and air conditioning

Hz	Hertz
INVC	Industrial Noise & Vibration Centre
IPPC	Integrated Pollution Prevention and Control
ISO	International Organization for Standardization
kHz	Kilohertz
L'_A	A-weighted sound pressure level at the protected ear
L_A	A-weighted sound pressure level(s)
LAeq	A-weighted equivalent sound level(s)
LAeq.T	Equivalent (or average) sound (or noise) level in dB over a period of time
LAmax	Maximum sound level(s)
L_C	C-weighted sound pressure level(s)
LCpeak	C-weighted peak sound level(s)
Lep,d	Daily personal noise exposure(s)
LOAEL	Lowest observed adverse effect level(s)
L_p	Sound pressure level(s)
L_{pA}	A-weighted emission sound pressure level(s)
Lr24hr	Noise level at the receiver over 24 h
LrD	Noise level at the receiver in the daytime
L_w	Sound power level [of a sound (or noise) in dB]
LwA	A-weighted sound power level(s)
N/m^2	Newton per square metre
NIHL	Noise-induced hearing loss

NIPTS	Noise-induced permanent threshold shift
NOEL	No observable effects level(s)
NPSE	Noise Policy Statement for England
Pa	Pascal(s)
PNR	Predicted noise-level reduction
PPE	Personal protective equipment e.g., hearing protectors
R	Sound reduction index
R_c	Room constant
rpm	Revolutions per minute
S	Slow-time weighting
SD	Standard deviation
SNR	Single number rating
SOAEL	Significant observed adverse effect level(s)
TTS	Temporary threshold shift
W	Watts
WHO	The World Health Organization
α	Sound absorption coefficient

Index

1/3 octave band, 6, 33, 223

A

A-weighted decibel(s) (dBA), 3-6, 10, 13-14, 16, 18, 34, 46, 53-54, 56-57, 59, 61, 70, 77, 94-97, 111, 114-115, 118, 124, 129-130, 136-140, 142-143, 150, 152, 156, 160, 165, 168, 170-171, 174-176, 178, 180-181, 185, 188, 191, 208-210, 213-214, 217, 219, 221-227, 230, 235-236, 238, 246-253, 255-258

A-weighted emission sound pressure level(s) (L_{pA}), 118, 188, 191

A-weighted equivalent sound level(s) (LAeq), 14-16, 35, 95-96, 111, 128, 152-156, 158-159, 175, 182, 236

A-weighted noise energy, 111

A-weighted sound energy, 14, 34

A-weighted sound power level(s) (LwA), 118, 188, 191

A-weighted sound pressure level(s) (L_A), 136, 138, 140

A-weighted sound pressure level at the protected ear (L'_A), 138, 140

Absorb(ed), 22, 26, 64, 120, 166

Absorbing, 8, 17, 22, 25-27, 33, 36, 43, 47, 56-61, 64-67, 70, 85, 94, 134, 157, 165, 170, 173, 176, 201-202, 205, 212

Absorption, 22-23, 25-28, 32, 36-37, 39, 42, 47, 53, 56, 61, 64, 66, 70, 85, 90, 94-95, 163, 166-168, 173-174, 177, 181-182, 185, 202-204, 209

    coefficient, 26, 28, 37, 61, 64, 66, 167

Absorptive, 47, 65-67, 75, 92, 209, 211, 221-222, 232-233

    silencer, 232-233

    splitter silencer, 65-66

Acoustic(s), 6, 8-9, 17, 20, 33, 35, 37, 45-47, 50, 53-59, 60-66, 68, 72-73, 78-79, 82, 85-87, 94-98, 101, 108, 110-111, 121, 131-132, 135, 150, 154-155, 157, 166, 174-175, 182-183, 185, 189-191, 194-199, 203, 205-206, 208, 211-212, 221, 224-225, 227-229, 232, 241, 245-246, 250, 255-256

    -absorbent lined shroud, 246

    absorption, 95, 166

    attenuator, 64

    cladding, 20, 35, 47, 54-55, 61

C

H

V

Milton Keynes UK
Ingram Content Group UK Ltd.
UKHW051947071024
449327UK00026B/2206